Game Theory - From Idea to Practice

Edited by Branislav Sobota

Published in London, United Kingdom

Game Theory - From Idea to Practice
http://dx.doi.org/10.5772/intechopen.102220
Edited by Branislav Sobota

Contributors
Ian M. Robinson, Mark Flanagan, Trevor C. Lipscombe, Adrian Northey, Matheus Cezarotto, Pamela Martinez, Barbara Chamberlin, Javier Vilamitjana, Julio Calleja-Gonzalez, Diego Marqués-Jiménez, Klaus P. P. Jantke, Hans-Holger Wache, Ronny Franke, David Ginat, Yasuo Nakayama, Corey Reutlinger, Branislav Sobota, Emília Pietriková, Merav Hayak, Orit Avidov-Ungar

First published in London, United Kingdom, 2023 by IntechOpen
IntechOpen is the global imprint of INTECHOPEN LIMITED, registered in England and Wales, registration number: 11086078, 5 Princes Gate Court, London, SW7 2QJ, United Kingdom

British Library Cataloguing-in-Publication Data
A catalogue record for this book is available from the British Library

Additional hard and PDF copies can be obtained from orders@intechopen.com

Game Theory - From Idea to Practice
Edited by Branislav Sobota
p. cm.
Print ISBN 978-1-83768-059-7
Online ISBN 978-1-83768-060-3
eBook (PDF) ISBN 978-1-83768-061-0

For EU product safety concerns:
IN TECH d.o.o., Prolaz Marije Krucifikse Kozulić 3, 51000 Rijeka, Croatia,
info@intechopen.com or visit our website at intechopen.com.

Meet the editor

Branislav Sobota is an associate professor in the Department of Computers and Informatics of the FEEI at the Technical University of Košice, Slovakia. He obtained his MSc (with honors) from the same institution in 1990, defended his Ph.D. in 1999 and a habilitation thesis in the field of virtual reality and computer graphics in 2008. His research interests focus in particular on virtual reality and related technologies, as well as computer graphics and parallel computing.

Contents

Preface IX

Chapter 1 1
Problem Solving of Mathematical Games
by David Ginat

Chapter 2 17
'Chance all' – A Simple 3D6 Dice Stopping Game to Explore Probability and Risk vs Reward
by Mark Flanagan,Trevor C. Lipscombe, Adrian Northey and Ian M. Robinson

Chapter 3 41
Developing Inclusive Games: Design Frameworks for Accessibility and Diversity
by Matheus Cezarotto, Pamela Martinez and Barbara Chamberlin

Chapter 4 61
Games and Social Reality
by Yasuo Nakayama

Chapter 5 83
Employing Games of Partial Information for Understanding Microaggressive Conflicts
by Corey Reutlinger

Chapter 6 99
Education and Games: Teachers' Professional Knowledge in Integrating Digital Games into Instruction in School
by Orit Avidov-Ungar and Merav Hayak

Chapter 7 113
Game Development and Testing in Education
by Emília Pietriková and Branislav Sobota

Chapter 8 141
Time Travel Gamification of Learning and Training: From Theoretical Concepts to Practical Applications
by Klaus P. Jantke, Hans-Holger Wache and Ronny Franke

Chapter 9 **165**
Methodological Approach in the Development of Specific Games
in Elite Soccer
by Javier Vilamitjana, Julio Calleja-Gonzalez and Diego Marqués-Jiménez

Preface

Games both as activities and as a basic educational tool are not only important from birth to death, but have been so from the beginnings of human society to the present day. The discipline of game theory deals mainly with types, description, algorithmization and strategies, but also the formalization of games. Indeed, games and the gamification of some human activities on various platforms have become an important part of our daily lives. Even philosopher and pedagogue J.A. Komensky spoke of "a school through play". Games in all shapes and forms contribute to the development of society. Today's young people will have played at least one game during their life, and as technologies develop, more people are playing games than ever before.

Modern game theory as a science is a very broad area. This book seeks to reveal part of what lies behind the "curtain" of game theory as a whole. It describes some game types, game theory tasks, the math background to game theory, game semantics, rule development, game goals or objectives, and the mechanics, design and development of the field, including computer games. The book includes, but is not limited to, a description of multiplayer and single-player games, cooperative and non-cooperative games, symmetric and asymmetric games, simultaneous games, turn-based games, complete and incomplete information games, scoring systems, and optimal play or winning strategies. Further topics include optimization of turns, time games and game complexity, games testing, games evaluation, games presentation, and the social, economic and business impact of games. Finally, there are the education and training aspects of gamification, or learning through playing.

I would like to express my sincere gratitude to all the authors and co-authors for their contributions. The successful completion of this book has been the result of cooperation between many individuals. I would especially like to thank IntechOpen Publishing Process Managers Marina Dušević and Dragan A. Miljak for their support.

Branislav Sobota
Technical University of Košice,
Košice, Slovakia

Chapter 1

Problem Solving of Mathematical Games

David Ginat

Abstract

Mathematical games are problems that involve algorithmic solutions. The solutions require recognition of hidden patterns and capitalization on these patterns. The natural tendency of many problem solvers is to devise algorithms without fully unfolding patterns. Such an approach lacks rigor and may lead to undesired outcomes. This chapter underlines a rigorous approach, of first focusing on the characteristics of a posed game and then developing its algorithmic solution. The solution development "goes" hand-in-hand with the realization of correctness. The approach is based on declarative observations, which capture the "what" of patterns prior to the "how" of game-strategy instructions. We illustrate the approach with colorful mathematical games of different characteristics and underline elements of solution processes, including creativity, problem-solving features, and mathematical notions.

Keywords: problem solving, Mathematical games, creativity, declarative and operational perspectives, invariance

1. Introduction

Mathematical games are mostly two-player games. Their specifications involve initial positions (states) and sets of rules for the players' moves. The two players play in alternating turns, and the game ends upon reaching a final position. The winner is usually the player who makes the last move. Most games are *complete information games*, in the sense that the game information is known at any given time to both players. The assumption is that each player always makes the best move available for her. Positions that lead to a win are *winning positions* for the player who moves next, and positions from which one will lose, regardless of her following moves, are *losing positions*. The player who moves first is *the first player*, and the other is *the second player*.

In most games, an *a priori* analysis of the game may "tell" which initial positions are winning positions and who will be the winner. The challenge for problem solvers is to employ an analysis that will reveal for each initial position whether it is a winning or a losing position and to offer a sound playing strategy for winning the game [1]. The winning strategy involves algorithmic instructions, based on unfolded underlying patterns. Recognition of hidden patterns and capitalization on patterns are the core of mathematical problem solving [2].

The primary inclination of many problem solvers is to invoke an *operational perspective* [3, 4], in which they focus on the instructions for the winner to follow (e.g., "if the position is such, then apply move-1, else apply move-2"). However, such a route may lead to erroneous playing strategies as well as partial argumentation of correctness. A preceding *declarative (assertional) perspective* is desirable. Such perspective focuses primarily on explicit identification and specification of declarative characteristics of the problem (game) at hand, before devising algorithmic operations [3–5]. It expresses thinking at the problem level [6] of a given problem and constitutes the patterns on which an operational solution will be based and argued to be correct. A problem-solving process that combines declarative and operative perspectives enables zooming in and out between abstraction levels—the higher patterns level and the lower concrete-operation level [7].

In this chapter, we demonstrate the latter with problem-solving processes that involve creativity for revealing underlying, hidden patterns. The creativity is expressed with flexible thinking and associations [8, 9]. The displayed processes do not always end with explicit algorithmic statements, but rather with insightful patterns, from which the reader may infer the algorithmic operations and realize their justification.

The next section displays solution processes of colorful two-player games. The games are simply stated and involve basic mathematics, yet their problem solving is not straightforward. We posed the games to mathematics/computer science (math/CS) junior and senior students, including some CS Olympiad trainees, and observed diverse problem-solving approaches. We elaborate on our experience and shed light on characteristics of student behaviors.

2. Operative and declarative game-solution routes

This section displays different problem-solving processes that illuminate the assets and importance of the declarative perspective. The section is divided into five sub-sections. Each sub-section illustrates, with one or more games, a feature that appears in game problem solving. The problem solving of some games involves more than one feature. Some features are general in mathematics and/or algorithmics, beyond problem solving of games.

In a displayed problem-solving process, we lead the reader through constructive and creative observations [2]. When the reader will follow a presented solution process, the train of thought may seem simple, since it is developed through suitable observations. Yet, in our experience, many problem solvers struggle. We use the conventions of calling the first player in all the games *Alice* and the second player *Beth*. In addition, we do not always indicate winning/losing positions; and in one game the last to move is not necessarily the winner.

2.1 Pairing

The first game involves basic geometry. The solution of the game is based on *a priori* pairing of game elements, such that when a player engages one of the elements of a pair, the response to that is to engage the element that was *a priori* paired with this element.

2.2 Closed shape on a grid

Alice and Beth play a game of connecting points in an $M \times N$ grid, in alternating turns. Each player, on her turn, connects two adjacent points, vertically or horizontally, which were not yet connected. Alice plays first. She connects points with red lines. Beth connects points with blue lines. Alice's goal is to obtain a closed polygon composed of *solely* red lines. Beth's goal is to prevent Alice from obtaining her goal. Would you prefer to be Alice or Beth in order to win the game?

Figure 1 shows a possible instantiation of the game, which ended in Alice's win. The last move in the game, made by Alice, was the drawing of a red line that "closed" the perimeter of the red polygon. (It could be any of the red lines of the perimeter.) Beth may win in another possible instantiation of the game, if she will manage to "interrupt" Alice so that there will be no red-perimeter polygon.

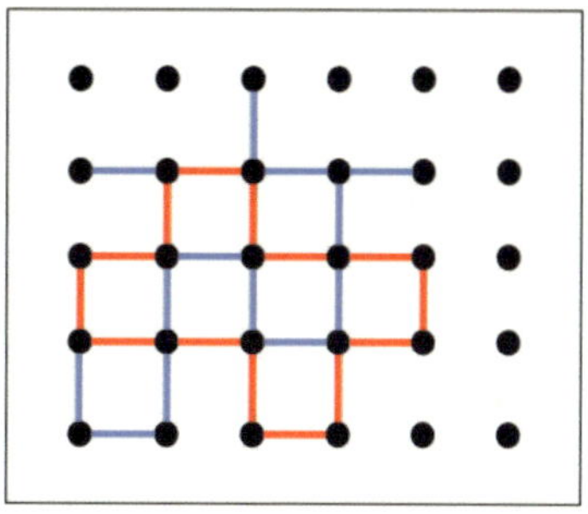

Figure 1.
A game that ended in Alice's win.

Problem solvers play the game with one another and offer various heuristic rules. Many feel that Beth may always win the game, but their strategies of line drawing are often vague and not rigorously justified. They attempt different strategies for "blocking" Alice but are unable to pinpoint a solid problem characteristic on which to capitalize. They experience difficulties with relating the game rules to a concrete property of "blocking." A creative, declarative perspective observation unfolds such a property.

- *Every polygon has a bottom-right corner, of a ⌟ shape, composed of a vertical line and a horizontal line.*

This simple property captures elegantly a characteristic of *every* polygon on the grid. It involves **pairing** of the two lines that compose the specified corner. A winning strategy for Beth is rather clear—for every new red line that Alice will draw, Beth will respond with its paired line (that will be blue), such that the two lines will compose a bottom-right corner. (Some boundary lines need no response.) The justification is

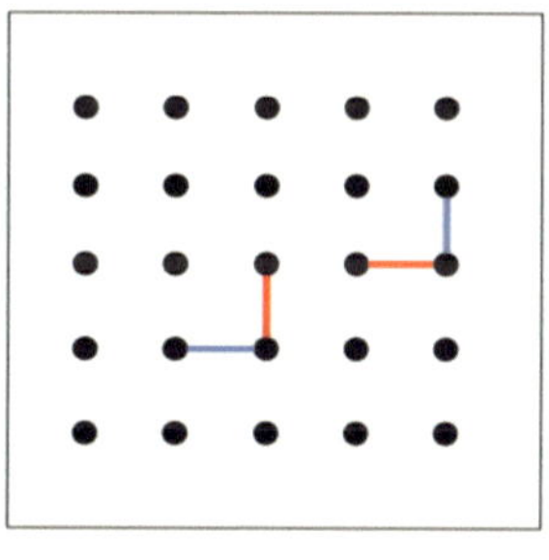

Figure 2.
Blue-line responses of Beth to Alice's red lines.

clear from the above characteristic. The two possible blue-line responses of Beth for a red line of Alice are displayed in **Figure 2**.

They key point in solving the game is that of relating a suitable "grid-polygon" property to the idea of "blocking." We noticed that such property may be tied to the playing feature of pairing (which will also appear in later games here). But just invoking pairing is insufficient. One has to examine polygon characteristics and identify a property to tie to pairing. Those who do not seek a corresponding polygon property do not reach a sound solution.

2.3 Invariance

The game below involves collection of numbers in a line. Unlike many mathematical games, it may end in a draw. The challenge is to devise a strategy that will lead to a win or a draw. It was posed as the first task of the 1996 International Olympiad in Informatics (IOI'96) [10]. The game solution is based on unfolding an appearance of the mathematical feature of invariance. Many problem solvers do not turn to this feature.

2.4 Left/Right Ends

Alice and Beth play a game of removing numbers from a line of 2 N numbers. Each player collects a number on her turn from one of the ends (a player may choose different ends in different turns). The removal of numbers ends when the line is empty. At this stage, each player sums the numbers she collected. Devise a strategy for Alice (the first player), which guarantees that her sum will be either larger than or equal to Beth's sum.

Many problem solvers follow an operative approach that focuses on *local* ends of the given line. They notice right away that it may be wrong to select the end number that is larger than the other end, since the inner value next to that chosen end may be large (and will become available to the opponent). They suggest looking at the *delta* of each end, which is the difference between the end number and the number next to it. They offer to select the end number from the end whose delta is larger; that is, select

7	8	9	5	2	2	8	10	4	6

Figure 3.
A game line of eight numbers.

the better "gained minus exposed." We illustrate this idea with the line of numbers in **Figure 3**.

Alice will compute the two "end deltas"—7–8 on the left end and 6–4 on the right end. Since −1 is smaller than 2, Alice will take the 6 from the right end. Beth will take any end number that she wishes, and Alice will calculate again the new "end deltas" in the same way for her next choice. And so on.

At first glance, this local strategy seems promising and leads to a win for many line examples. (Does it lead to a win in the line below?) However, its local focus on the line ends, and not on the whole line, may lead to a loss. Falsifying examples are not immediate, but one may find lines of length smaller than 10 for which this strategy will fail. Some problem solvers do not notice the latter and "argue" correctness with partial, heuristic argumentation. Others who do notice the difficulty try to "patch" it in diverse unsuccessful ways.

One should seek a sound declarative observation on which to capitalize. The natural tendency may be to focus on the values of the given line numbers. But a creative, flexible thinking may lead to the examination of *locations* of the numbers in the original line. The line includes an even amount of numbers. This yields an illuminating observation when we look at the **parities of the locations of the original line**. (Half of the locations are odd, and half are even.)

- *In* **every move of Alice**, *she will have a choice between two numbers, whose parities of locations in the original line are different (one even and one odd).*

- *In* **every move of Beth**, *she will have a choice between two numbers, whose parities of locations in the original line are equal.*

We examine these observations with the line of numbers in **Figure 3**. If, for example, Alice will remove the 6, then Beth will have to choose between 7 and 4, which were originally in odd locations. Regardless of which number Beth will choose, Alice will receive a line whose ends were in locations of different parities in the original line.

At this point, a playing strategy can be devised for Alice. She may *force* a game in which she will collect all the numbers that are originally in the odd locations or all the numbers that are originally in the even locations. Thus, in the beginning of the game, she will see which sum is larger—that of the numbers in the odd locations or that of

7	8	9	5	2	2	8	10	4	6

Figure 4.
Line numbers colored according to the parities of their locations.

the numbers in the even locations—and decide accordingly. The numbers of these sums are illustrated in red and blue in **Figure 4**.

The creative aspect of this strategy stems from examining locations, *without* looking at the values in these locations (in the beginning of the analysis). This demonstrates the asset of turning to a declarative perspective, in trying to obtain an insight *into the problem*, prior to the development of any algorithmic solution.

Many problem solvers are "fixated" on values throughout their problem-solving process and miss this perspective. Such a phenomenon is typical of those who turn immediately to an operative perspective, seeking an algorithmic solution right away, without carefully analyzing first the problem at hand.

The above observations are specified in an **invariance** manner, in the sense that they relate to **every move** of a player, throughout the game. Invariance is a powerful means in both mathematics and computer science, for "capturing" properties of recurring states (e.g., [11]). As such, it is most relevant for games, as a means for characterizing recurring positions throughout a game. We shall see an additional example of invariance in the next games, when formulating a winning strategy that forces a sequence of losing positions on the loser of a game.

The displayed strategy involves **pre-processing**, which occurs here when the game data is provided and before making the first move. Pre-processing is a useful and powerful feature in algorithm design [11].

The presented strategy guarantees that Alice will not lose the game, but there may be initial lines for which she will end in a draw when she can actually win. A computational approach, employing dynamic programming [11], will yield a maximal sum for Alice in each game. The interested reader is welcome to program the latter. It will be based on computing the maximal sum that can be obtained by the winner in a separate game for *every* sub-sequence of the original line.

2.5 State-space reduction

In many games, the general state space may be very large. However, the winner may direct the game, so it will be played in a very particular path, and many legal states (positions) will be out of reach. The path will involve a limited number of winning and losing positions. One occurrence of this phenomenon may be reached by disabling the usage of game pieces. The following problem solving illustrates this phenomenon, with a variant of the game of domino.

2.6 Domino line

Alice and Beth play domino with a pile of N×(N + 1)/2 stones, of the pairs <1,1>, <1,2 > up to <N-1,N>, <N,N>. The order of the two integers in a stone does not matter; that is, the single stone <K,L > may be rotated and become <L,K>. Each player *sees* all the stones in the pile throughout the game. At the beginning of the game, Alice picks from the pile any stone that she wishes and starts a line of stones by putting the first stone. From this point on, each player picks a stone from the remaining pile and *lengthens the line*, to the left or to the right (as she wishes), by putting her chosen stone according to the domino rules (neighboring ends of adjacent stones match). **Figure 5** displays a possible game state after five moves. The game ends when there are no stones in the pile with which one can

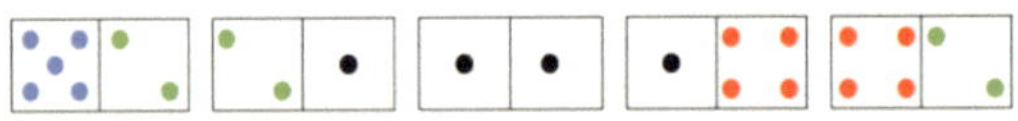

Figure 5.
A game state after five moves.

lengthen the line. The player who makes the last move wins. Devise a strategy for winning the game.

At first glance, problem solvers get the impression that this is a simple game to analyze. Yet, many struggle, sometimes considerably. Some problem solvers seek a winning strategy by using all the game stones. Others try a table for keeping track of used stones. And some represent the game with a graph and try to devise an algorithm for expanding a path in the graph. All these attempts yield partial operational schemes, which are cumbersome and/or erroneous.

The solution of this game may be reached in two stages of declarative observations. A common theme in many games is an ordered reply to every opponent move. This was explicit in the first game that we presented, with the notion of pairing. This notion could be applied here if each domino stone was doubled. In such a case, Beth could win by responding to every move of Alice with a paired stone.

Unfortunately, we do not have such pairs of stones in our game. Yet, in a creative first step, we may temporarily simplify the game, by leaving out stones that may look more "problematic"—the "doubles" stones, of the form <K,K>. And we may create pairs with **some** of the remaining stones, in an original way, according to the following observation.

- *In a game with no doubles, one may pair some of the stones, using the integers 1 and 2, such that for every X > 2, the two stones < 1,X > and < 2,X > will be a pair.*

This pairing does not involve all the stones of the game. But we may direct the game so that only stones with 1 or 2 will actually participate in the game. This can be done according to the following observation.

- *In a game with no doubles, if Alice will start the game by putting the stone < 1,2>, then Alice may* **force the game** *to be played only with stones of the form < 1,X > and < 2,X > (X > 2), as Alice may reply to Beth's stone with its paired stone.* **After every move** *of Alice, each end of the line will be either 1 or 2.*

Figure 6.
The game state after three moves, of forcing only stones with 1, 2, or both.

Figure 6 displays a line of three stones reached after three moves according to the above idea. Alice started with the left-most <1,2 > stone, Beth put <2,4>, and Alice replied with <4,1 > – the stone paired with <2,4 >.

At this point, we may "return" the doubles to the game. The addition of the doubles may seem problematic at first glance, as doubles do not have paired stones. But with some flexibility, of using the idea of a "long stone," we may overcome this challenge.

- *In the original game, with doubles, it is possible to create a* **"long stone"** *from a sequence of three stones that will serve as the stone < 1,2 > in the previous observation. If Alice will start with the stone < 1,1>, Beth will follow with the stone < 1,Z>, and Alice will respond with the stone < Z,Z>; the sequence < 1,1 > <1,Z > <Z,Z > may be viewed as a "long <1,Z> stone," and from this point on, the integers 1 and Z will serve as 1 and 2 in the previous observation.*

Figure 7 illustrates the above "long stone" idea, where the number 5 is Z.

The above creative observation yields a winning strategy for Alice, in a game in which only stones with 1 and Z will be used.

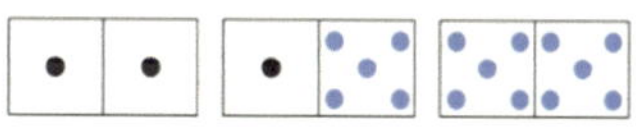

Figure 7.
A "long" <1,5> stone.

The strategy correctness is rather clear, based on the creative idea of the "long <1,Z> stone" and the notion of pairing. Notice that this winning strategy **narrows** the game considerably. It **forces** the game to be played with a small set of stones. The path in which the game is played **alternates between winning positions and losing positions.** In particular, Alice leads Beth in each of her moves to a losing position, which is characterized by line ends that are either 1 or Z. This narrows considerably Beth's options of choosing a valid stone, and for any stone that she will choose, Alice has a reply. Notice the **invariance** feature of the line after each of Alice's moves.

The problem-solving process progressed gradually, through three observations, obtained from flexible, creative ideas. These ideas express **divergent thinking**. Although the end result is simple and clear, progress could not occur without such thinking. Problem solvers who are not invoking the notions of pairing and invariance and are less creative struggle here with unfruitful heuristic attempts.

2.7 In spite of incomplete information

Sometimes one may devise a strategy that is only based on the game information provided in the game specification. While the game information will be incomplete during the game, one may still win by knowing only the rules that the opponent obeys. This is illustrated with the following game.

2.8 Wall of doors

Alice and Beth play a game with a wall of N doors. The doors are numbered 1 to N. Alice walks along the wall and opens and closes doors. Beth moves behind the wall from door to door. Alice *does not know* Beth's location throughout the game. When the game starts, Beth stands behind one of the doors. Alice may open any door that she wishes in order to see if Beth is behind. If Beth is there, the game ends with Alice's win. Otherwise, Alice closes the door, and Beth moves to an *adjacent* door, on the right or left (as she chooses; though if she is near an end door, then she has only one choice). Alice does not see Beth's move. Then, Alice opens a door again, which can be the same door or any other door. The game continues for at most 5 N opening attempts. If Alice finds Beth, then she wins; otherwise Beth wins. Would you like to be Alice or Beth?

Figure 8 displays a game state in which the red-hair Alice is about to open door 3.

Figure 8.
A game state in which Alice is in front of door 3 and Beth is behind door 6.

Beth is behind door 6. After Alice will realize that Beth is not behind door 3, Alice will close that door, and Beth will move to door 5 or door 7. Alice will then try to open any other door, or the same door, and the game will continue accordingly.

As we may see, Beth's location is hidden from Alice throughout the game, including the beginning. The question is whether Alice can always find Beth in spite of her limited information. Problem solvers attempt various operational methods for Alice. Many conjecture that Alice should start from one of the ends, move in one direction, and sometimes try to open doors twice (or more). Their attempts are usually not clearly phrased and not displayed with sound argumentation of correctness.

A declarative perspective may illuminate problem characteristics on which to capitalize. We present three observations, which are unfolded gradually, in trying to understand the occurrences that lead to opening a door behind which Beth is standing. The first observation is trivial. Yet in our experience, it is not that naturally invoked, as problem solvers are "busy" with operational ideas for Alice, rather than with observations.

- *If Alice will open a door that is* **two doors away** *from Beth and then they will both move toward each other, then Alice will find Beth.*
- *So is the case if Alice will open a door that is of an* **even distance** *from Beth, and they will move toward each other, while Alice is opening* every *door on the way.*
- *And so is the case if their initial distance is even when Alice will open a door and will then move toward Beth, opening every door on the way, and* **Beth will move as she likes** *(including reaching the end of the wall and moving back).*

The last assertion implies that if Alice will first open a door in one of the wall ends and then open every door on the way to the other end, then she will find Beth if their initial distance is even. However, this "one pass" will not end successfully if their initial distance is odd. The odd distance between them will be kept all along. The following observation offers a way to change the parity of the distance between them.

- *Opening a door twice in a row changes the parity of the distance between the girls.*

Thus, if Alice will reach the other end of the wall without finding Beth, she will infer that their initial distance was odd. Alice will open the end door twice and find Beth on her way back to the first door. All in all, Alice will not open more than 2 N + 1 doors.

The gradual declarative observations paved the way to the elegant solution. The underlying mathematical element was **parity of the distance** between the two girls. We may recall the relevance of parity (of locations) in the Left/Right Ends game. As in the previous games, here too one may formulate an invariant for describing and arguing about the correctness of the devised strategy.

2.9 Natural number properties

Many games involve natural numbers. One such game is the well-known game of Nim [1]. The winning strategies in these games capitalize on properties of numbers and involve the features of pairing and invariance. Some also relate to the notions of parity and/or symmetry. We display below three games with such features.

2.10 Powers of 2

Alice and Beth play a game with 10 piles of tokens. The first pile is with one (2^0) token, the second is with 2^1 tokens, the third is with 2^2 tokens, and so on. The largest pile contains 2^9 tokens. Each player, on her turn, selects 5 of the piles (as she wishes) and removes one token from each of the selected piles. The game ends when no move can be played (too few non-empty files). Devise a strategy for winning the game.

Problem solvers who turn to the operational approach in trying to devise a strategy experience difficulties with the sole token in the smallest pile. A declarative approach will first focus on patterns of powers of 2. The following basic property may be relevant.

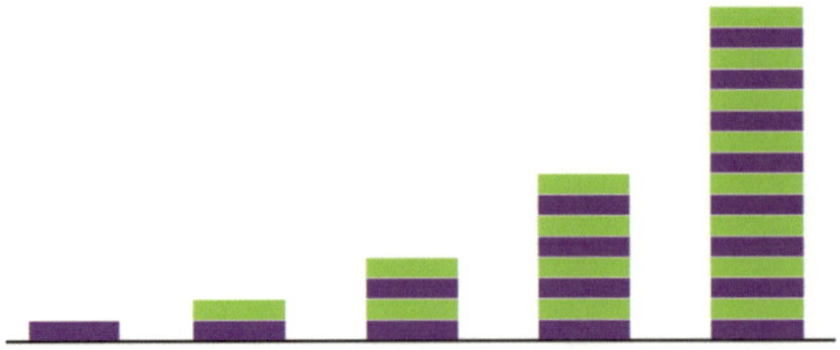

Figure 9.
The smaller five piles of the ten piles.

- *Every power of 2 is greater (by 1) than the sum of all the powers of 2 that are smaller.*

Figure 9 displays the first five piles. One may visually notice the observation above about the relation between each pile and those on its left.

This property may be elegantly tied to our game, with the following observation.

- *The game will end when the 6 smallest piles will be emptied, and at this point, none of the 4 largest piles will be empty.*

The latter observation is justified by the recognition that even if each of the four largest piles will be selected in every move of the game, one token will still be removed from one of the six smaller piles. The observation yields a simple useful idea that addresses the difficulty that stems from the sole token in the smallest pile.

- *The 4 largest piles may be used as a "***bank***" from which tokens may be removed, in order to reach an invariant property with the 6 smaller piles.*

Alice may win the game by removing in her first move the single token of the smallest pile plus a token from each of the four largest piles and then keeping the following **invariant**.

- *The number of tokens in each of the 6 smaller piles will be even after each move (of Alice).*

Alice will keep the above invariant by duplicating each of Beth's moves. The amount of 0 tokens in each of the six smallest piles will be reached only by Alice. In retrospect, the simple property of powers of 2, plus the idea of using the larger piles as a "bank," paved the way to the winning strategy.

2.11 Co-Primes

Alice and Beth play a game with a line of N successive positive integers, N > 11, from K to K + N-1. Each player, on her turn, removes an integer from the line. After N-2 turns, two integers will remain. If these two integers are co-prime, then Alice wins; otherwise, Beth wins. For which initial sequence of integers will Alice win, and for which will Beth win? Devise a strategy for winning the game.

Upon approaching the problem, one may wonder whether the values of N and K are important. In our experience, quite a few problem solvers do not examine N and

K at first glance, and seek a winning strategy for Alice, by focusing on removal of integers that have many divisors. They attempt some heuristic strategies but do not advance much. Meanwhile, some notice that every second integer in the initial line is an even number and change their initial train of thought, this time seeking a winning strategy for Beth.

The main theme in these hasty, heuristic attempts is their operational approach, based on intuitive, partial observations. The more competent problem solvers seek underlying characteristics. They first analyze the given problem and wonder about properties of co-primes. Obviously, a prime number is a co-prime with any other number. But it is unclear whether such a characteristic may be useful. It may be better to seek another co-prime characteristic. Further examination yields a simple and elegant observation.

- *Every two successive integers are co-primes.*

This observation implies a simple winning strategy for Alice when N is odd, which will be based on the following line pattern.

- *Every line of 2L successive integers may be divided into* **pairs** *of successive integers, which are co-primes.*

The latter pattern implies that for the case of odd N, Alice may divide the initial line into adjacent pairs of successive integers, except for the left (or right) end. She will first remove this end and then respond to each of Beth's moves by removing the integer paired with the one just removed by Beth.

So the problem is solved for the cases of an odd N, with the **pairing** feature. May this strategy be relevant also for the cases of an even N? Not quite. The player who plays first may not pair all the integers remaining after her first move. When N is even, the number of even integers in the line equals the number of odd integers. We notice the following.

- *Alice* **must** *remove an even integer in* each *of her (N/2)-1 moves.*

If Alice will not do so, Beth may lead the game to two even integers in the end. But if Alice will solely focus on the even numbers, Beth may leave two odd integers that are multiples of 3 if the initial line is long enough. What is long enough?

- *Every line of more than 11 successive integers includes at least two odd integers that are multiples of 3.*

Since the initial line length is larger than 11, Beth will win the game when N is even. All in all, the declarative perspective gradually paved the way to the winning strategies with a series of corresponding observations.

2.12 Powers of primes

Alice and Beth play a game with two piles of tokens. Each player, on her turn, removes from one of the piles a number of tokens that is a power of a prime (the power 0 is also legal). For example, from a pile of 15 tokens, one may remove in one turn any number of tokens apart from 6, 10, 12, 14, and 15. The game ends when both piles are empty. Devise a strategy for winning the game.

As with previous games, quite a few problem solvers attempt various operational ideas of removing tokens. Some recall a two-pile variant of the game of Nim and attempt symmetry, trying to equalize the amount of tokens in the piles. They realize that this idea may not be applied in some initial positions. Others focus on emptying one pile first (e.g., the smaller) and then proceeding to the second. They feel that their ideas lack rigor.

Gradual progress combined with declarative observations offers a fruitful and sound result. As a first step, we may solve the case of a single pile. Upon examining game scenarios with very small piles, we may notice the following.

- *In a game with a single pile, the amount of 6 tokens is the smallest amount from which the game will not end in one move, since 6 is the multiplication of two different primes. Thus, 6 is a losing position.*

Further examination of game scenarios with larger piles reveals that in a game with a single pile, the amounts of 12 and 18 are also losing positions (for the player who is about to play). A corresponding generalization may be observed. We specify it in two (equivalent) ways—with the indication of losing positions and with invariance.

- *In a game with a single pile, the losing positions are ones with token amounts that are multiples of 6. The winner should keep the following* **invariant**: *after each of her moves, the amount of tokens in the pile will be a multiple of 6.*

The initial positions in which the number of tokens is not a multiple of 6 are winning positions, as in each of them the first to play may lead to the above invariant in one step. At this stage, we may move to the second step—the original game.

Figure 10.
Piles that cannot be equalized in one move.

For some initial positions, it is impossible to equalize the amounts of tokens in the piles in one move. This is the case in the example of **Figure 10**, with piles of token amounts of 3 and 9.

Perhaps it is possible to obtain some variant of equal amounts. We noticed the importance of multiples of 6. With some flexible, creative thinking, we reach the following declarative observations related to the notion of remainder.

- *If the* remainders *of division by 6, of the token amounts in both piles, are equal, then the next move will yield different remainders, and the move that follows may yield equal remainders.*

- *In a game with two piles, the losing positions are ones in which the* remainders *of division by 6, of the token amounts in both piles, are equal. The winner should keep the following* **invariant***: after each of her moves, the remainders of division by 6, of the token amounts in both piles, will be equal.*

All in all, for the initial positions in which the difference between the amounts of tokens in the piles is not a multiple of 6, Alice will win. In all the other initial positions, Beth will win. The path to the final observation was gradual, and as with previous games, declarative observations and creativity paved the way.

3. Conclusion

We illustrated the essential role of the declarative perspective in problem solving of mathematical games. This perspective leads the problem solver to first examine characteristics of the problem at hand and then capitalize on them in devising an algorithmic winning strategy. Such a problem-solving process yields not only the algorithmic solution but also its justification. The focus on problem characteristics expresses the highest level (level 4) of abstraction in the ladder of **abstraction levels** during algorithm design [6]. In particular, this level is higher than the program level (level 2) of this ladder, which corresponds to the operational perspective that focuses on the actual algorithmic instructions.

The illustrated declarative perspective in the game solutions expresses suitable employment of the component of *control* in Schoenfeld's cognitive model of problem solving [2]. This component "directs and monitors" progress in the process and combines the invocations of relevant resources and heuristics as well as justifications. Invoked *resources* involve mathematical properties such as polygon characteristics, list locations, parity, modular arithmetic, and number characteristics. They also involve game features such as pairing, invariance, and pre-processing. Invoked *heuristics* involve problem decomposition, modification, simplification, and more. Suitable invocations and applications of these elements paved the way to elegant game solutions that were displayed here.

A primary additional component of the cognitive model is that of *beliefs*, including beliefs in the way to approach a problem and progress and evaluate the outcome. The main difference between more competent and less competent problem solvers stems from the demonstration of more proper and less proper control and beliefs [2]. Many problem solvers that we described followed an operational perspective that was insufficient for obtaining sound outcomes, in terms of suitable algorithmic solutions and rigorous justifications. The competent problem solvers sought sound underlying patterns prior to, or together with the operative outcomes.

The different practices also affected creative behaviors. Those who expressed divergent, flexible thinking while seeking patterns reached constructive ideas on which they capitalized, while those who were focused right away on the final algorithmic outcomes demonstrated limited points of view and fixation. One must develop **awareness** of the important role of the elements described and illustrated here. Such development may occur with realization of the differences between the suitable and less suitable, or unsuitable, routes for progress.

Author details

David Ginat
Tel-Aviv University, Tel-Aviv, Israel

*Address all correspondence to: ginat@tauex.tau.ac.il

References

[1] Berlekamp ER, Conway JH, Guy RK. Winning ways for your mathematical plays. 2018

[2] Schoenfeld AH. Learning to think mathematically: Problem solving, metacognition, and sense making in mathematics (Reprint). Journal of Education. 2016;**196**(2):1-38

[3] Denning PJ, editor. A debate on teaching computing science. Communications of the ACM. 1989;**32**(12):1397-1414

[4] Ginat D. Mathematical operators and ways of reasoning. The Mathematical Gazette. 2005;**89**(514):7-14

[5] Ginat D. Hasty design, futile patching and the elaboration of rigor. ACM SIGCSE Bulletin. 2007;**39**(3):161-165

[6] Perrenet J, Groote JF, Kaasenbrood E. Exploring students' understanding of the concept of algorithm: Levels of abstraction. ACM SIGCSE Bulletin. 2005;**37**(3):64-68

[7] Wing JM. Computational thinking. Communications of the ACM. 2006 Mar;**49**(3):33-35

[8] Torrance EP, Rockenstein ZL. Styles of thinking and creativity. Learning Strategies and Learning Styles. 1988:275-290

[9] Ginat D. Learning from wrong and creative algorithm design. ACM SIGCSE Bulletin. 2008;**40**(1):26-30

[10] https://ioinformatics.org/page/ioi-1996/22

[11] Cormen TH, Leiserson CE, Rivest RL, Stein C. Introduction to Algorithms. Cambridge, Massachusetts: MIT Press; 2022

Chapter 2

'Chance all' – A Simple 3D6 Dice Stopping Game to Explore Probability and Risk vs Reward

Mark Flanagan, Trevor C. Lipscombe, Adrian Northey and Ian M. Robinson

Abstract

'Chance all' is a simple 3D6 dice game that explores a player's attitude to risk vs reward. Strategies for playing the game are explored ranging from zero risk to more complex forms of risk, based on an appreciation of the odds; those strategies more likely to win are identified. In addition, the game may be an indirect measure of an individual's bias towards risk vs reward and how that bias alters through the game as the likelihood of winning and losing changes. It can be used as a simple introductory teaching tool for the Gaussian distribution to examine chance and probability, in evolution and computing science, together with psychological aspects of gameplay.

Keywords: dice game, stopping game, probability, Pi, normal distribution, winning strategies, risk and reward

1. Introduction

'Chance all' is a new dice game that uses three six-sided dice (3D6). The game proceeds by rounds. Initially, each player has a score of 0. Each player then throws 3D6 and adds the scores from each die, giving their cast, which must lie between 3 and 18. They are free to keep this number for the value of this round (i.e. stop) or to roll again (i.e. Chance all). If the new cast has a value equal to or less than the original cast, the score for this turn is 0, and the next player takes their turn. If the new cast exceeds the value of the original cast, the player adds the scores for the first and second casts together. Again, the player is invited to stop or to continue to 'Chance all', with the score being added to the previous value provided it is greater than the last cast. If it is equal to or less than the last cast, then the value for this round is 0, and the next player has their turn. This cycle of pushing their luck continues until the player feels ready to stop, or loses all scores this round. The next player repeats the procedure for their turn until all players have had their go in the round. In subsequent rounds, the value from the preceding round is the starting value and either zero or the value for the turn is added on, giving a cumulative score round by round. The game continues until the first player exceeds a cumulative score of 100 and is declared the winner. If, in the same round, more than one player exceeds 100, the player with the highest cumulative total wins. A flowchart for the game is presented in Appendix 1. The starting player in each round can be rotated in a clockwise manner to avoid any bias. This simple game has a rich complexity that is explored in the sections below.

2. Investigating the results from 3D6

As the game involves rolling 3D6 dice and adding the scores to give a total value for the cast (the throw), the probability distribution involved can be investigated by a number of methods. First, one can simply count the outcomes for three dice. The first record of this in literature appears to be a reference dating to 1678, due to Strode [1]. The total number of throws that yield a given 3D6 cast from 3 to 18 is shown in **Figure 1**, and counting each 3D6 result gives values that match Strode.

An analytical way to determine the probability distribution for rolling n 1D6 dice (where n is the number of 1D6 dice to be thrown) is via the expansion of

$$\left(x + x^2 + x^3 + x^4 + x^5 + x^6\right)^n \tag{1}$$

Each coefficient A in the series expansion below (i.e. $A.x^m$) represents the number of throws that gives the cast value m, where $n \leq m \leq 6n$. In the case of 3 dice, n = 3 and the expansion is

$$\left(x + x^2 + x^3 + x^4 + x^5 + x^6\right)^3 = 1.x^3 + 3.x^4 + 6.x^5 + 10.x^6 + 15.x^7 + 21.x^8 + 25.x^9$$
$$+27.x^{10} + 27.x^{11} + 25.x^{12} + 21.x^{13} + 15.x^{14} + 10.x^{15} + 6.x^{16} + 3.x^{17} + 1.x^{18} \tag{2}$$

Strode's value, and counting from **Figure 1** for a cast of 8 with 3D6, gives 21 different ways of achieving this total. This can also be found by inspecting Eq. (2) for the coefficient A in the series where m=8 i.e. $21.\times^8$, which is in agreement. Further mathematical methods to derive the coefficients for 3D6, such as combinatorial or recursive techniques, are explored more fully by McShane and Ratliff [2], and these give the same results above. For brevity, they are not reproduced here.

The counting method given in [1], from **Figure 1**, and from the probability generating function method [2] above all agree. The number of throws for each cast from 3 to 18 for 3D6 is given in **Table 1**, together with the probability for each throw expressed as both a fraction A/216 from the definition above in Eq. (2), and as a percentage to 1 dp. Finally, in the last column, the probability of a throw being greater than a given cast is indicated. This is the sum of all possible outcomes higher than the given value, and it is expressed both as a reduced fraction and as a percentage to 1 significant figure, calculated in **Table 1**. As an example, a cast of 8 has 21 possible throws, and it has a probability of 21/216 (9.7%). The chance of getting a cast greater than 8 in a subsequent throw is 20/27 (74.1%).

Figure 2 visually resembles a Gaussian distribution, an observation explored more fully later by statistical analysis. **Figure 3** shows the probability of throws exceeding a cast against a given 3D6 cast.

Knowledge of the data in **Table 2** and **Figures 2** and **3** clearly allows a player to make more informed choices on whether to stop or to continue to 'Chance all' each

1

6	8	9	10	11	12	13
5	7	8	9	10	11	12
4	6	7	8	9	10	11
3	5	6	7	8	9	10
2	4	5	6	7	8	9
1	3	4	5	6	7	8
	1	**2**	**3**	**4**	**5**	**6**

2

6	9	10	11	12	13	14
5	8	9	10	11	12	13
4	7	8	9	10	11	12
3	6	7	8	9	10	11
2	5	6	7	8	9	10
1	4	5	6	7	8	9
	1	**2**	**3**	**4**	**5**	**6**

3

6	10	11	12	13	14	15
5	9	10	11	12	13	14
4	8	9	10	11	12	13
3	7	8	9	10	11	12
2	6	7	8	9	10	11
1	5	6	7	8	9	10
	1	**2**	**3**	**4**	**5**	**6**

4

6	11	12	13	14	15	16
5	10	11	12	13	14	15
4	9	10	11	12	13	14
3	8	9	10	11	12	13
2	7	8	9	10	11	12
1	6	7	8	9	10	11
	1	**2**	**3**	**4**	**5**	**6**

5

6	12	13	14	15	16	17
5	11	12	13	14	15	16
4	10	11	12	13	14	15
3	9	10	11	12	13	14
2	8	9	10	11	12	13
1	7	8	9	10	11	12
	1	**2**	**3**	**4**	**5**	**6**

6

6	13	14	15	16	17	18
5	12	13	14	15	16	17
4	11	12	13	14	15	16
3	10	11	12	13	14	15
2	9	10	11	12	13	14
1	8	9	10	11	12	13
	1	**2**	**3**	**4**	**5**	**6**

Figure 1.
The 216 different chances from 3D6. The number in bold shows the individual 1D6 rolls, with the resulting 3D6 cast in normal text.

3D6 cast (range 3–18)	Throws corresponding to cast	Probability (%) of throw	Fraction and probability (%) of throw greater than cast indicated
3	1	1/216 (0.5)	> 3 215/216 (99.5)
4	3	3/216 (1.4)	> 4 53/54 (98.1)
5	6	6/216 (2.8)	> 5 103/108 (95.4)
6	10	10/216 (4.6)	> 6 49/54 (90.7)
7	15	15/216 (6.9)	> 7 181/216 (83.8)
8	21	21/216 (9.7)	> 8 20/27 (74.1)
9	25	25/216 (11.6)	> 9 5/8 (68.5)
10	27	27/216 (12.5)	> 10 1/2 (50.0)
11	27	27/216 (12.5)	> 11 3/8 (37.5)
12	25	25/216 (11.6)	> 12 7/27 (25.9)
13	21	21/216 (9.7)	> 13 35/216 (16.2)
14	15	15/216 (6.9)	> 14 5/54 (9.3)
15	10	10/216 (4.6)	> 15 5/108 (4.6)
16	6	6/216 (2.8)	> 16 1/54 (1.8)
17	3	3/216 (1.4)	> 17 1/216 (0.5)
18	1	1/216 (0.5)	

Table 1.
3D6 casts, the total number of throws that yield this value, the fraction and probability of the throw (% to 1dp). Finally the fraction and probability of the cast being exceeded (% to 1dp).

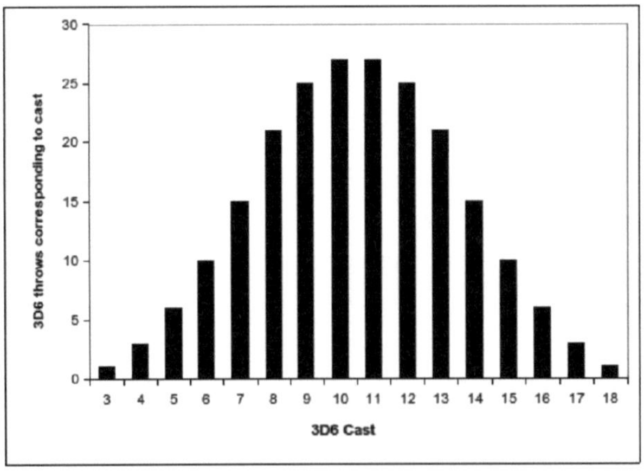

Figure 2.
Histogram of 3D6 casts vs throws, by plotting the data from ***Table 1.***

round, based on their most recent cast. 3D6 casts are discrete whole numbers ranging from 3 to 18, rather than being a continuous function. Exploring the distribution seen in **Figure 2** with statistical methods is instructive. The Gaussian distribution is a continuous function given by

Round	Zero risk (zr)	Single risk (sr)	Double risk (dr)	Fixed risk (fr) Bt = 10	Tapering risk (tr) Bt = 13	Ramping risk (rr) Bt = 7
1	6	11,**8**	11,**10**	10	7, 7	11
2	9	4,12	18, **12**	12	5,10,15	6, 9
3	9	8, 10	16, 5	11	8,11	4,9,14
4	9	16,**12**	12,13,18	12	9	7,9,**4**
5	12	10,**6**	14,5	10	8	9,14
6	13	5,7	10,15,7	8,10	10	7,12,**9**
7	10	12,**11**	11,7	13	7	7,13,15

Table 2.
3D6 casts from run 8 in the tournament showing each cast. Those shown in bold have a cast equal to or less than the preceding cast and consequently, score zero for that round.

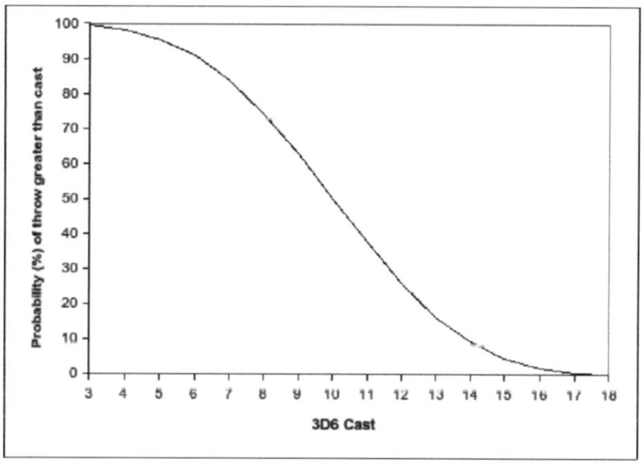

Figure 3.
Histogram of 3D6 casts vs probability of throws exceeding the indicated cast value, by plotting column 1 against column 4 from ***Table 1***.

$$f(x) = \frac{1}{\sigma\sqrt{2\pi}} exp^{-\frac{1}{2}\left(\frac{x-\mu}{\sigma}\right)^2} \quad (3)$$

where μ is the mean of the distribution and σ is the standard deviation. One way to test whether a dataset conforms to the Gaussian distribution is to perform an Anderson-Darling statistical test [3]. The data are plotted against a theoretical normal distribution in such a way that the points should form an approximately straight line when plotted using a normal probability plot. The data in **Table 1** were tested using the Anderson-Darling function using the statistical program Minitab v19 [4], and the results are plotted in **Figure 4**.

The red line indicates the probability expected from the data if it conformed to the Gaussian distribution derived from the data shown as dots, both on a normal probability plot. As can be seen in **Figure 4** the Gaussian distribution is a good approximation to the data between casts 5 to 16, with the low and high tails lying off the predicted normal probability. The P value for the Anderson-Darling test (0.009) confirms the data is not Gaussian; nonetheless, apart from the low (<5) and high casts (>16),

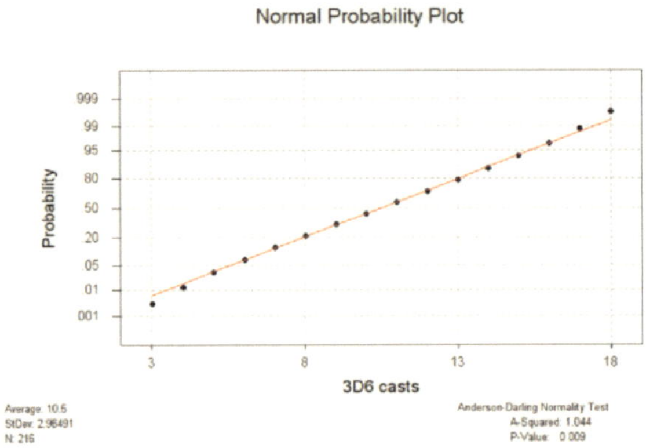

Figure 4.
Anderson-Darling test performed on throws corresponding to casts from 3D6.

it conforms well. The mean value is 10.50, the standard deviation is 2.96, the skewness is 0.00 and the kurtosis is −0.40, all to 2dp. The skewness value confirms the data is equally distributed about the mean. The kurtosis value [5] confirms the data is platykurtic, meaning it has 'thinner' tails than expected from a Gaussian distribution, given that 3D6 casts are discrete and cannot physically be less than 3 or greater than 18. The Gaussian distribution is continuous and given σ = 3.0, the $\pm$ 3σ levels at a probability of 0.3% on either side of the mean would require dice rolls of 1.5 and 19.5, which is not possible with 3D6. Nonetheless, 3D6 serves as a readily available means of generating 'Gaussian like' data in a classroom setting, which should serve as introductory teaching and provide a good basis for the game, assuming the dice are 'fair' – inasmuch as the result of each dice is random in nature, with the Chi-squared statistical test as a means of detecting bias in the distribution in repeated dice rolls. In addition to physical dice, electronic dice rollers are readily available, such as Google Dice [6], which the authors have checked against the Chi-squared test, with no bias detected. Electronic dice rollers use a pseudorandom number generator (prng), such as xorshift128+ [7] or the Mersenne Twister [8], to generate the dice results. References based on the Diehard Tests are presented in Appendix 2 to check the randomness of prng's. Alternatively, the decimal number sequence of π provides a ready source of random numbers and has passed numerous statistical checks [9–13] to this effect. It is readily available [14], and converting it to a suitable form for dice casts used in the game is discussed in Appendix 2.

3. Probability and psychology or risk vs reward

'Chance all' can be considered to be a stopping problem in a gambling sequence to gain a benefit each round that leads to winning the overall game. Thus, knowledge of the probabilities in the game as outlined in Section 2 is helpful for a smart player, whereas a naive player will play the game intuitively. Set against the random process in the game, which is 'Gaussian-like,' and the inherent risks this causes is the psychology of an individuals sense of risk vs reward.

Stopping games have their own literature [15–20] and are used in gambling, economics and control theory. The solutions to each game or system tend to be unique to the game being played and concentrate on defining an algorithm that seeks to balance the reward vs the risk or probability of continuing to gamble. Being mathematical solutions, these tend to remain constant throughout the game.

Games are played by people rather than algorithms, and as a counterpoint, there are papers that examine the game from a psychological perspective, as perception of risk changes depending on how far away or close you are to winning, or where you appear in the pack of players. Notably, Liu et al. [21] described an interesting concept. Repurposing their lemma 1 to 'Chance all', for each round of the game R, there is a unique benefit threshold, Bt, such that the player will stop if Bt is matched or exceeded, or roll the dice if the cast is less than Bt. In addition, they note the benefit threshold may decrease as the current winning player gets closer to the winning level, as the player seeks to consolidate gains and their position. Likewise, a player who is losing may well accept more risk and increase Bt as the game progresses in order to overtake the player in the lead. Consequently, a player may change Bt throughout the game depending on their position in the pack, and the process of the game may dominate their sense of risk vs reward. This may well also be altered by age and experience. Younger students may be affected by playing with dice themselves, their sense of dice 'fairness', and their belief in non-physical parameters. Watson and Mortiz [22] quote earlier studies suggesting that some students place agency on God, fate or mental powers to determine dice outcomes, rather than accepting dice casts as a random process. Adults are known to be prone to the gambler's fallacy [23, 24] a well-understood phenomenon that can be described as apophenia. Some players in a game believe that if a particular event occurred more frequently than normal during the past it is less likely to happen in the future. This is not true in a game where the outcome is random and is perhaps due to cognitive bias and a failure to appreciate the logic of random behaviour. Silverman et al. [25] noted that in 200 consecutive coin tosses runs of at least 6 heads or 6 tails are to be expected. In π,the string of digits 12345 occurs after the decimal place 49,702. Thus low-odds events are to be anticipated in a random long sequence. There is also the propensity to hot hand [26], where people who have correctly guessed a short sequence of random draws in a game believe they can continue to do so. Hot hand is thought to be caused by the illusion of control and that individuals can influence or have pre-knowledge over random events. Clearly, people of all ages are complex in their beliefs and decision-making. Perhaps simple algorithmic approaches that are consistently applied against people as players might fare well in 'Chance all'. Strategies that could be employed every round but which are independent of personal judgement and involve a benefit threshold Bt might include:

3.1 Zero risk

3D6 is thrown and the cast is accepted regardless of the outcome, and no further throw is made that round. In this sense, Bt is 3 for every round and the cast is accepted. Such an approach is likely to be suboptimal if a very low cast is made given the high probability of exceeding it. It is however a zero-risk approach.

3.2 Single risk

3D6 is thrown with a given cast. Regardless of this value, another throw is made. If the second cast is equal to or less than the first cast the score for that round is zero. If not, the player stops, adds the two casts together and no further throw is made this round. One could say that Bt = 19 using this method. Clearly, such an approach is sub-optimal if 18 is cast in the first round, guaranteeing failure in the second.

3.3 Double risk

3D6 is thrown with a given cast. Regardless of this value, another throw is made. If this second cast is equal to or less than the first cast, the score for that round is zero. If not, the player makes another throw. If this third cast is equal to or less than the second cast the score for that round is zero. If not, add the three casts together and no further throw will be made this round. Again, one could say that Bt = 19 using this method. This approach is sub-optimal if 18 is cast in either the first or second round, guaranteeing failure in subsequent throws.

3.4 Fixed risk

A unique critical benefit level, Bt is defined before the game starts such that $3 \leq Bt \leq 17$, and is held constant throughout the game. 3D6 is thrown with a given cast and is compared to the defined Bt. If the cast matches or exceeds Bt the player stops. If the cast is less than Bt, another throw is made with a given cast. If the second cast is equal to or less than the first cast the score for that round is zero. If not, the player stops, adds the two casts together and no further throw will be made this round. Bt can be gauged from a careful appreciation of Section 2. If Bt were set to 10, then 50% of casts would be expected to exceed the value, making it equally likely to succeed or fail.

3.5 Tapering-down risk

A unique critical benefit level, Bt is defined before the game starts such that $3 \leq Bt \leq 17$, and is steadily reduced throughout the game by using Bt - R, where R is the round in the game, starting at 1 and increasing by 1 each round. 3D6 is thrown with a given cast and is compared to the defined Bt - R for that round. If the cast matches or exceeds Bt - R, the player stops. If the cast is less than Bt - R, another throw is made with a given cast. If the second cast is equal to or less than the first cast the score for that round is zero. If not, the player adds the two casts together. Further casts can be made this round in a similar manner until the threshold for stopping is reached or the cast is equal to or less than the previous cast in the process when the score for that round is zero. If Bt were set to 13, then the threshold for round 1 would be 13-1 = 12 and 26% of casts would be expected to exceed the value, making it likely that another cast is made at the beginning. By the time round 5 was reached, the threshold would be 13-5 = 8, and 74% of casts would be expected to exceed the value, making it less likely that more casts are made at this stage. This mimics being eager for reward and accepting the risks initially, but reducing the chances of failure as the game goes on at the expense of likely gains. This matches Liu et al.'s [21] observations on human behaviour for a winning player.

3.6 Ramping-up risk

A unique critical benefit level, Bt is defined before the game starts such that $3 \leq Bt \leq 17$, and is steadily increased throughout the game by using Bt + R, where R is the round in the game, starting at 1 and increasing by 1 each round. 3D6 is thrown with a given cast and is compared to the defined Bt + R for that round. If the cast matches or exceeds Bt + R, the player stops. If the cast is less than Bt + R, another throw is made with a given cast. If the second cast is equal to or less than the first cast, the score for that round is zero. If not, the player adds the two casts together. Further casts can be made this round in a similar manner until the threshold for stopping is reached or the cast is equal to or less than the previous cast when the score for that round is zero. If Bt were set to 7, then the threshold in round 1 would be 7+1 = 8 and 74% of casts would be expected to exceed the value, making it less likely that another cast is made at the beginning. By the time round 5 was reached, the threshold would be 7+5 = 12, making it more likely that more casts are made at this stage. This mimics being risk averse initially, then increasing the chances of increasing gain at the expense of failure as the game goes on. This matches Liu et al.'s [21] observations on human behaviour for a losing player eager to win.

To test these strategies, a tournament was made for strategies A-F using the suggested Bt levels for the fixed, tapering and ramping risk approaches. The confidence interval (CI) for a value depends upon the number of measurements made

$$CI = \underline{x} \pm z\,\sigma/\sqrt{n} \tag{4}$$

where $\underline{x}$ is the mean, z the confidence interval level, σ the standard deviation and n is the number of measurements. Beyond a certain number of measurements, there is reduced benefit in repeated sampling, so the number of trials was fixed at 20, and the confidence interval was set at the 95% level, giving z = 1.96.

The value for all the 3D6 dice rolls in the game and the final scores for each strategy were recorded. An example game is shown below, using Google roll dice [6] as the means of generating casts. For each game, the strategies were run together, and the game stopped when a winning strategy was found.

Table 3 is plotted below in **Figure 5** and shows the game sequence for this run.

As can be seen, ramping risk won this game, with fixed risk second, tapered risk third, zero risk fourth and single and double risk in second to last and last place respectively. This run demonstrates the strengths and weaknesses of each of the defined strategies. Zero risk steadily gains, but at a slower rate than riskier strategies and fails to win. Single risk failed to score in rounds 1, 4, 5 & 7. We leave it to the reader to decide whether they would have continued after the first cast, but clearly

Round	Zero risk (zr)	Single risk (sr)	Double risk (dr)	Fixed risk (fr) Bt = 10	Tapering risk (tr) Bt = 13	Ramping risk (rr) Bt = 7
1	6	0	0	10	0	11
2	15	16	0	22	30	26
3	24	34	0	33	49	53
4	33	34	43	45	58	53
5	45	34	43	55	66	76
6	58	46	43	73	76	76
7	68	46	43	86	83	111

Table 3.
Cumulative sum of 3D6 casts from run 8 in the tournament for each round, by adding the results from each round shown in ***Table 2*** *above. Ramping risk won this run by exceeding the target of 100.*

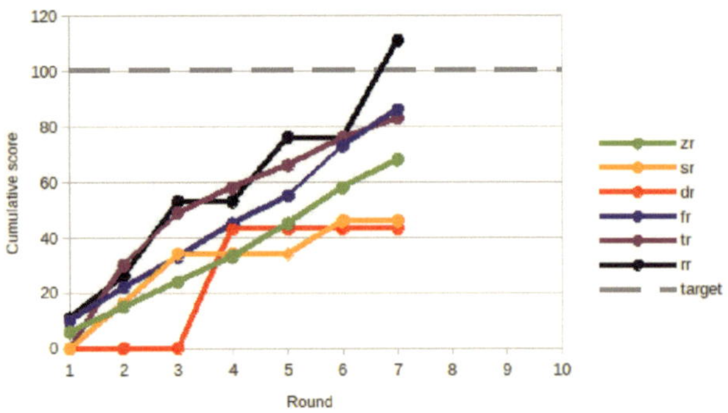

Figure 5.
Cumulative scores from run 8 in the tournament for zero risk, zr (green); single risk, sr (orange); double risk, dr (red); fixed risk, fr (blue); tapering risk, tr (purple) and ramping risk, rr (black). Data plotted from ***Table 3****.*

automatically rolling again can fail. Double risk failed to score in rounds 1–3, and 4–7, and scored the lowest overall of all the strategies. Fixed risk, with Bt = 10 always scored and outperformed the three strategies above. Setting Bt = 10 meant on only one occasion a second cast was required in round 6, which paid off. Tapering risk failed to score in the first round, but scored in all subsequent ones, with round 2 being especially successful. The final cast in this sequence exceeded the Bt for that round (13-2 = 11) thus stopping the sequence. It finished just above fixed risk in this sequence. Ramping risk failed to score in rounds 4 & 6 but won the game. In round 6, three strategies were close in cumulative score (73 to 76) and yet the ramping risk won because the Bt threshold had risen (8+6) = 14, and the final cast exceeded the stopping value. This run is illustrative of the various methods, but the true test for each strategy comes from repeated runs where the odds even out. **Table 4** shows the final scores when the game ended after 20 runs, together with the mean and standard deviations, and the number of times each strategy won.

The mean and standard deviations from **Table 4** allow the calculation of the confidence interval for the mean from Eq. (4) showing the range at ±95% confidence levels. Also shown is the number of times the strategy was winning in the tournament. The most successful strategies in terms of wins and average score per round across all the tournaments were ramping risk (35% wins & 12.3), tapering risk (30% wins & 12.7) and fixed risk (10% wins & 12.8), followed by zero risk (5% wins & 10.4), then single (20% wins & 9.6) and double risk (0% wins & 4.5). Ramping risk won the runs most times, just ahead of tapering risk. The confidence intervals show there is no significant difference for the top three strategies (ramping, tapering and fixed risks) in terms of mean final score and standard deviation, and this is confirmed by t-testing (P = 0.52) [27]. There is a significant difference between these three highest scoring strategies to the next highest strategies (zero risk and single risk) as determined by a t-test. Finally, these strategies are significantly different to the poorest strategy (double risk) as determined by a t-test (P = 0.0004). All the top winning strategies are willing to roll the dice twice or more based on the odds from the initial dice roll, which distinguishes them from double risk, which throws regardless of the odds. Ramping risk has a rising Bt throughout the game, which pushes the number of double or triple rolls on as the game is played ('Who Dares Wins') and this was clearly successful a number of times with the strategy overtaking others at the end. Conversely, tapering risk takes its chances early then swings to a conservative strategy as the round

Run	Zero risk	Single risk	Double risk	Fixed risk	Tapering risk	Ramping risk
1	76	95	0	102	84	**108**
2	74	62	20	91	89	**114**
3	73	**100**	73	84	63	73
4	68	108	67	67	**110**	78
5	46	61	42	**100**	99	88
6	61	**104**	33	57	69	65
7	92	63	88	99	**100**	94
8	68	46	43	86	83	**111**
9	83	85	33	79	84	**107**
10	82	**101**	0	87	91	79
11	56	42	0	98	**105**	67
12	98	**122**	94	100	112	97
13	52	59	0	87	**111**	80
14	58	55	0	94	54	**113**
15	**112**	79	0	72	94	65
16	74	77	0	98	**109**	60
17	85	16	33	77	**110**	57
18	74	26	44	92	103	**108**
19	86	35	37	106	91	**109**
20	60	44	31	**106**	74	73
$\underline{x}$ (σ)	74.0 (16.6)	72.3 (29.5)	31.7 (32.1)	87.2 (12.7)	92.8 (17.4)	87.3 (19.8)
CI @ 95%	66.8 / 81.0	52.0 /86.0	18.6 / 45.2	83.3 / 94.9	84.3 /99.2	78.6 /96.0
Wins	1	4	0	2	6	7

Table 4.
Final scores for strategies A–F after 20 runs in the trial. Scores in bold indicate the winning total for that trial. $\underline{x}$ *(σ) are the mean and standard deviation in brackets.*

progresses, tending to almost a zero-risk approach at the end. If there are enough early gains to consolidate a lead in some games it may win, although early failures condemn this strategy to lose that run. Fixed risk does well in terms of average score, but wins less often against the other two riskier strategies. It outcompetes zero risk, which fails to accept the risk of rolling twice, and only succeeded once (5% of runs). Single risk had a highly variable track record, winning 20% of the games but sometimes failing to score well at all. This contrasts strongly with fixed risk, which has a similar strategy but is based on a careful appreciation of the odds. Nonetheless, single risk won 20% of games and fixed risk 10% of games. To examine this further, Appendix 3 presents a probability analysis for the expected average score per round for zero risk (Bt = 3) and fixed risk ($4 \leq Bt \leq 18$). This explains why fixed risk, if Bt is chosen carefully, is a superior strategy to zero risk, which must yield on average a cast per turn of 10.5. The analysis suggests that Bt = 11 is optimal, yielding an average score per round of 13.3 to 1dp. It also suggests why ramping- and tapering-risk strategies work well since their

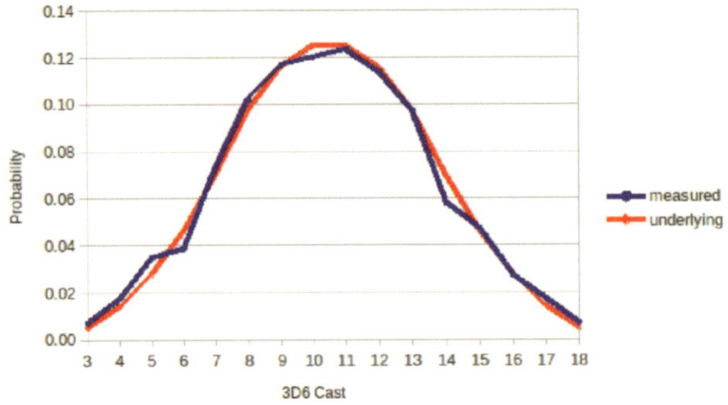

Figure 6.
Probability for 1433 3D6 casts using Roll Dice Google app [6] in the game, compared to the underlying counted distribution for the 216 combinations of 3D6 given in ***Table 1****. This passes the Chi-squared test.*

Bt values straddle the maximum central portion on average score per cast. From this, it is possible to select algorithmic strategies to compete in tournaments with people, with zero risk and fixed risk at Bt = 11 suggested. These observations are only valid if the process of rolling 3D6 to obtain casts is random and is thus unbiased (**Figure 6**). To confirm this, every 3D6 cast in the tournament (1433 in total) was counted for the number of times 3–18 occurred, allowing their probability distribution to be calculated and compared to the underlying distribution shown in **Table 1**.

Owing to the nature of randomness, one should never expect a random number sequence to exactly match a theoretical calculation. These data, though, suggest that Roll Dice [6] by Google is an acceptable means of generating random dice rolls and suitable for the game. It also validates that the repeated tests of the 20 runs of the game are unlikely to be biased and so the conclusions regarding the game strategies are sound.

4. Adaptation to an educational setting and trials by players

In terms of education, the game can be used in support of the mathematics curricula on elementary probability and statistics. One simple way to do this in a class of 30 students is to divide into 6 groups of 5 students and play the game without teaching any of the underlying mathematical structures. Each group records all the individual 3D6 casts and tracks the cumulative scores in the game. Once the final round has occurred and a winner is found, the group tabulates a histogram of their 3D6 casts and compares it to the underlying distribution shown in **Table 1** and **Figure 2**. It typically takes 8 rounds or more to complete the game, so this is likely to be at least 40 casts, with more likely using double casts. This is some way short of the 216 possible casts expected from 3D6 to give the full distribution. For example, histogramming the data in **Table 2** gives the graphs shown in **Figure 7**.

Each group is asked how the histogram compares to the expected 3D6 pattern and whether the 3D6 casts are a fair representation, given that the chance of a 3 or 18 cast is low. Each of the six groups in the class then pools their casts, resulting in a deeper data pool of some 240 to 300 casts. This histogram is then compared to the underlying distribution. After this stage, teaching about the expected probabilities in the game can help alter perceptions, and the students repeat playing the game. After each group plays again, they histogram their new 3D6 casts and add these to the first set for comparison

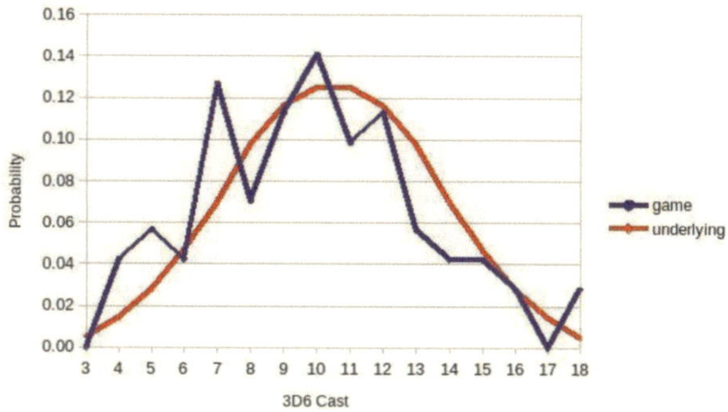

Figure 7.
Probability for 71 3D6 casts from ***Table 2*** *using Roll Dice Google app [6] in the game, compared to the underlying counted distribution for the 216 combinations of 3D6 given in* ***Table 1****. A t-test reveals no significant difference between the mean (std dev) of 9.90 (2.50) to the underlying mean (std dev) of 10.50 (2.96). This passes the Chi-squared test.*

with the fundamental distribution, with about 120 casts available. Pooling all the groups' data pools together and histogramming should result in an experimental set of data closer to the expected shape, more akin to **Figure 6**. It is also instructive to question students whether they changed their perception of the game having been taught about the underlying probability and the nature of risk vs reward.

The game can also be used to explore inheritance, variation and evolution in the biology curriculum using genetic algorithms [28], and in the computing curriculum as an example of a search algorithm. Suppose we define a single Bt value for each of ten rounds to complete the game, with each unique Bt created by rolling a 3D6 and using the cast as the value for that round such that $3 \leq Bt \leq 18$. Having done this 10 times, the string of Bt values represents a genetic algorithm, with the value for each Bt in the round representing the 'allele' for the gene, and its position is the location within the 10-step 'chromosome'. The whole sequence gives the 'genotype', with the sequence result in the game representing its fitness i.e. its 'phenotype' against a 'natural selection' process (i.e. the 'Chance all' game). By randomly creating say 4 such genotypes, the game is played, and the top two winning genotypes are selected to go through, creating new 'offspring' by the processes of recombination and mutation, before entering competition again with the previous top two genotypes. The process is repeated and the students can see the process of evolution occur over a number of different rounds. To illustrate this the following example is given.

First, four genotypes each with unique chromosomes were created via 3D6 casts giving the Bt value for each round in the 1st generation. Note there is nothing unique to these sequences.

Run sequence	1	2	3	4
1	9	10	12	15
2	11	11	9	16
3	8	10	12	14
4	8	8	16	7
5	10	13	10	15
6	6	8	13	10

7	8	7	10	7
8	7	8	12	12
9	17	11	5	9
10	15	12	8	17

One might expect 'genotype 4' to struggle in the game given the very high Bt thresholds in the early rounds, created by random chance given the expected average scores associated with Bt levels discussed in Appendix 3. The game was played using each of these Bt values each round to determine whether to continue to roll if the cast was beneath Bt or to stop if the cast matched or exceeded Bt. The individual scores in each round were:

Run sequence	**1**	**2**	**3**	**4**
1	6,14	4,11	9,11,**11**	8,**8**
2	9,**6**	9,5	9	12,**6**
3	7,13	9,11	7,8,12	10,12,**12**
4	10	12	16	9
5	13	6,12,**8**	16	15
6	12	13	14	10
7	12	12	13	8,7
8	13	10	10,**10**	13

Scores that failed the test of 'Chance all' are shown in **bold**. This led to final cumulative scores of genotype 1 = 100, genotype 2 = 82, genotype 3 = 95 and genotype 4 = 47. Since the selection test criteria of 'Chance all' were met after 8 rounds (or trials of the genotype), the last two 'chromosomes' of all genotypes were untested and could be described as redundant in this game, and latent in the process until a game that lasts 10 rounds is played. The first (genotype 1) and second (genotype 3) placed genetic algorithms in the game are judged to have passed the fitness for survival of the 'phenotypes' from their 'genotypes'. As expected, genotype 4 failed to win against the natural selection test of 'Chance all', due to the failure of the first three 'genes' to score: It went extinct, along with genotype 2. Genotypes 1 and 3 were selected to continue to 'evolve'. Rolling a 1d6 and using 1–3 by selecting the top 5 'genes', or 4–6 the bottom five 'genes' in the phenotype for recombination allows the 'children' to be created from their parents. In this example, a 3 was thrown, which resulted in the top five chromosomes being exchanged. Each of these five was then tested for mutation, with a 1D6 roll of 1 decreasing the Bt level by 1, 2–5 keeping them the same, and a roll of 6 increasing the Bt level by 1. The new genotypes or children, 5 & 6 are shown alongside the survivors from the first round, their parents, 1 & 3 in the second generation of the process. The children 5 & 6 have swapped the top five chromosomes shown in blue over with the other parent but kept the last 5 the same. The top five alleles underwent mutation by the process described above, with some increasing or decreasing in value by 1. These are shown in the following tabular runs by **bold** numbers, and unchanged values in normal font.

Run sequence	1	3	5	6
1	9	12	*12*	*9*
2	11	9	*9*	*11*
3	8	12	*12*	***9***
4	8	16	***17***	***9***
5	10	10	***9***	*10*
6	6	13	6	13
7	8	10	8	10
8	7	12	7	12
9	17	5	17	5
10	15	8	15	8

These 4 genotypes were then entered into the game as the 2nd generation, and the top two were selected by the test of the game mimicking natural selection as before. Cumulative scores for the 2nd generation after 7 rounds were genotype 1 = 110, genotype 3 = 72, genotype 5 = 100 and genotype 6 = 89, and so 'genotypes' 1 & 5 were selected as parents to the 3rd generation in the game. Recombination this time resulted in the bottom 5 'chromosomes' being exchanged, and after the mutation stage the next generation were

Run sequence	1	5	7	8
1	9	12	9	12
2	11	9	11	9
3	8	12	8	12
4	8	17	8	17
5	10	9	10	9
6	6	6	*6*	***7***
7	8	8	*7*	***9***
8	7	7	*7*	***6***
9	17	17	*17*	***16***
10	15	15	***16***	*15*

Note that the last five 'genes' for genotypes 1 & 5 are identical, but the process of mutation has altered them in their 'children', genotypes 7 & 8. Again, the test continues with the cumulative scores for the 3rd generation after 7 rounds being genotype 1 = 45, genotype 5 = 75, genotype 7 = 96 and genotype 8 = 103, and so 'genotypes' 7 & 8 were selected as parents to the 4th generation in the game. Note that no original genotype has now survived, and the 4th generation has evolved away from the 1st generation, but traces of the same gene sequences persist. The process of the game can continue for as long as required to teach the process of inheritance, variation, and evolution. One way of examining the effect of a change in climate on evolution in biology using this method would be to split a teaching class into groups to explore 'Chance all' by genetic algorithms as explained, but allow one of the groups to add +1 to each 3D6 roll mimicking a 'hotter climate' and another group to subtract −1 to each 3D6 roll mimicking a 'colder climate', and see the results each group come up with over a suitable number of generations. This would demonstrate the effect of local adaptation to conditions, with the 'hotter climate' 3D6+1 expected to evolve increased values of Bt in the genotype, and the 'colder climate' 3D6-1 expected to evolve decreased values of Bt in the genotype by the same processes of recombination and mutation to reflect the additional dice rolls. Roll dice [6] supports adding a modifier via a ±toggle to the basic cast on the right hand side of the dice option.

In terms of teaching computing, the genetic algorithm approach allows for 16 separate values for Bt for each 'allele' or position in the sequence, giving a search space in the order of 10^{16} possible combinations with 10 run positions. The genetic algorithm approach explores the possible optimal solutions relatively cheaply with no understanding of the underlying complexity of the mathematics in the game, but yet delivers viable solutions, and so is quick and cheap to implement.

Our tests of playing the game as individuals playing naively without an underpinning strategy tend to support the observations of Liu et al. [21] in that the desire to keep pushing your luck decreases for a current winning player as they get closer to the winning level of 100. Likewise losing players accept more risks as the game progresses in an attempt to try to overtake the leading player. This occurs in a game with open and symmetric information where all players can see the scores as the game progresses. Our experience of playing suggests the psychological aspect could be explored in further studies, including a measure of an individual's sense of risk vs reward. Contrasting play under open and symmetric information (all players see each other's casts), to asymmetric information (one player sees all casts, but the others know only their own) and closed and symmetric information (players only know only their own) would be interesting as the following method suggests. There are 4 players in each team, Red 1 to 4 and Blue 1 to 4, with Red 4 and Blue 4 following a zero risk methodical approach by rolling 3D6 once per round and no more (in effect dummy hands). All other players are free to pursue their own strategy.

The game proceeds in stages with the following matrix of information access and sharing.

Information Access \ Information Sharing	Symmetric	Asymmetric
Open	Stage 1 Red vs Blue	Stage 2 Blue Stage 3 Red
Closed	Stage 4 Red vs Blue	Stage 2 Red Stage 3 Blue

In stage 1, the game is played open (all dice rolls are declared by all players in public) and symmetric so the information is equal. The target score is set for 100. In stage 2, the game is played both open (all dice rolls by the Blue players are declared in public) and closed (all dice rolls by the Red players are hidden), leading to asymmetry in information. The target score is reduced to 90. In stage 3, the game is played both open (all dice rolls by the Red players are declared in public) and closed (all dice rolls by the Blue players are hidden), leading to asymmetry in information. The target score is reduced to 80. In stage 4, the game is played closed (all dice rolls are hidden by all players) and symmetric so the lack of information is equal to all parties. The target score is set to 70. The game ends after 4 stages. After each stage, one player is eliminated from the game by following these additional rules.

1. If the Red team wins, the lowest placed Blue player in that stage is eliminated (except 3, so the next weakest).

2. If the Blue team wins, the lowest placed Red player in that stage is eliminated (except 3, so the next weakest)

3. If a player wins the stage, they can't be eliminated in the next stage (i.e., a pass).

Everyone will want to win, and no one wants to be last in their group for fear of the group losing and being eliminated, and this threat on the weakest player in the losing group each stage keeps the pressure on to succeed. The information access and sharing and the reduction in target scores stage by stage create additional pressures for the players to explore. In the latter two modes of play, players declare if they have met the victory conditions at the end of each round. These alternatives to open play have the effect of making players uncertain of their ranking in the game as it progresses, and our experience is it tends to make players more risk-tolerant, for fear of losing. Controlled competitions with careful selection of the players using a 2 level factorial designed experiment for the variables of age and gender might be used to see if there is a bias in the game.

5. Conclusions

'Chance all' is a simple game to play with a surprising degree of mathematical structure. In addition, the game lends itself well to supporting teaching the curriculum in mathematics directly, and in biology and computing through the adoption of a genetic algorithm approach to playing the game. As such, it tests players' sense of risk vs reward and has a psychology to it that matches well with previous observations in similar games and settings. We have explored the game in an algorithmic manner and suggested optimal strategies based on the underlying probability described in Appendix 3. These can be tested against real players as optimised strategies, with Bt = 11 being suggested.

Chess computer programs work because the game is played by logic. 'Chance all' is a game based on random chance with known probabilities. Can an algorithm be created to beat people's intuition in an uncertain game? Further studies may give clearer answers to this question.

Different forms of the game are of course possible, based on the dice used. nDS dice have n = number of dice, and S = sides to each dice. 3D6 is the minimum that gives a

Gaussian-like distribution, for the smallest number of casts making the game as simple as possible, which is especially useful when considering its applicability to education.

'Chance all' takes moments to learn, yet its mathematical complexity and subtlety pose problems that provide fruitful avenues for further research, rather like other well-known mathematically rich games, such as Go, backgammon, and chess.

Appendix 1 Flowchart for 'Chance all'

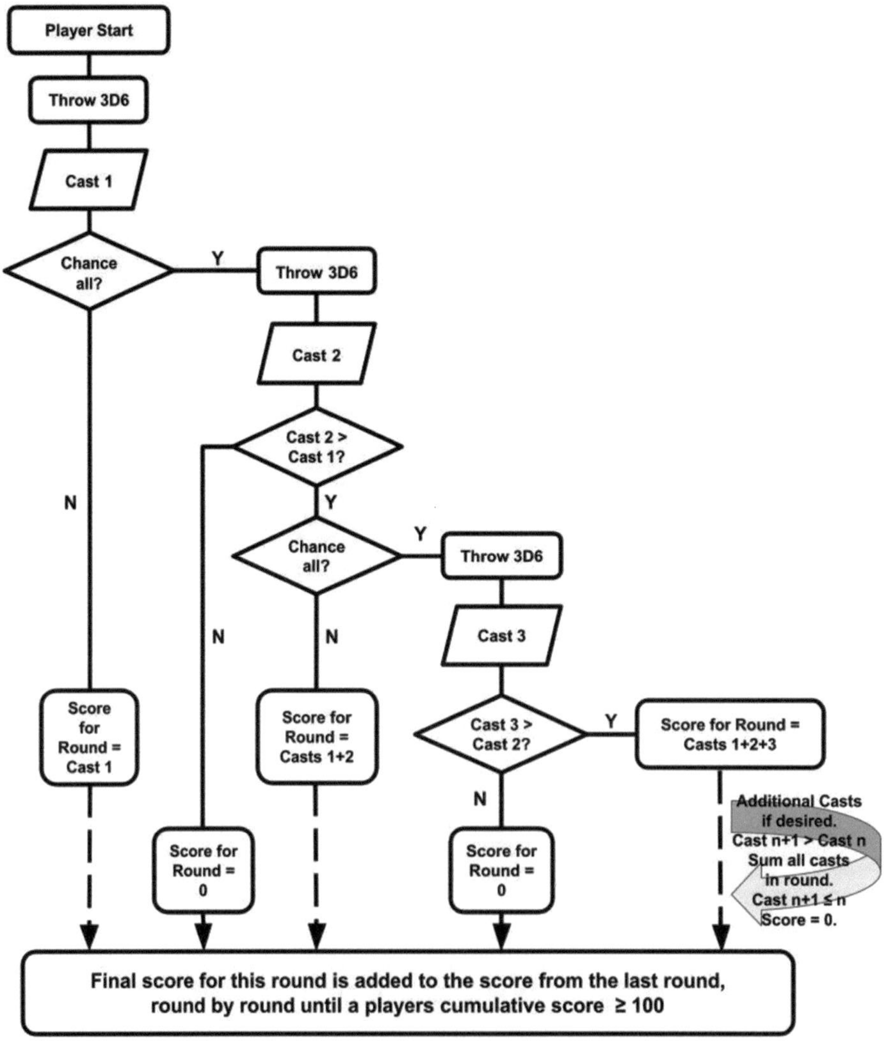

Players use the flowchart above for each round to determine their score, which can be either the first cast alone or comprised of multiple throws of 3D6 providing the last cast exceeds the previous cast, in which case the summed casts in the round are used. If the last cast is the same or less than the previous cast, then the score is 0 for that round. The final score for this round is added to the score from the last round, and this pattern continues round by round until the first player exceeds a cumulative score of

100 and is declared the winner. If, in the same round, more than one player exceeds 100, the player with the highest cumulative total wins.

Appendix 2 Converting π to a source of random dice rolls

The decimal number sequence of π provides a ready source of random numbers and has passed numerous statistical checks [8–12] to this effect. routines. Johnson and Leeming [11] found that π achieved higher randomness ratings than 100,000-digit runs from random number generators. π is readily available in a suitable form for use [14]. One can sample the string by selecting consecutive 5 numbers and using 00000 $\leq$ **1** $\leq$ 16666, 16667 $\leq$ **2** $\leq$ 33333, 33334$\leq$ **3** $\leq$ 50000, 50001 $\leq$ **4** $\leq$ 66667, 66668 $\leq$ **5** $\leq$ 83333, and 83334 $\leq$ **6** $\leq$ 99999, where the number in black represents the cast from 1-6. For example, 141592653589793 are the first 15 numbers after 3. in π, giving 14159 (**1**), 26535 (**2**), 89793 (**6**), and a 3D6 cast of 9. The results of this method are random, although the sequence is deterministic. After sampling 100,000 digits of π after 3. and breaking them into 5 number strings, converting these to dice rolls as above 20,000 individual 1D6 casts were calculated, and then summed into 6,679 consecutive 3D6 casts (**Figure 8**) [29]. The probability distribution can be calculated for these casts and compared to the underlying distribution from 3D6 outlines in **Table 1**.

The sequence can be intercepted at any point to generate a random sequence of dice rolls for use within the game using the string of digits after an arbitrary set position. As observed by G Marsaglia, the originator of the 'Diehard' tests [30] for randomness, '*The digits in the expansion of irrationals such as π, seem to behave as though they were the output of a sequence of independent identically distributed (iid) random variables*' [31].

Appendix 3 Probability for average scores in 'Chance all' for a given Bt

The analysis begins by selecting a Bt, such that 3 $\leq Bt \leq$ 18. If the first cast is s, then if s $\geq$ Bt, you will score the single roll total:

$$Single = \sum_{s=Bt}^{18} sp(s) = \sum_{s=3}^{18} sp(s) - \sum_{s=3}^{Bt-1} sp(s) = 10.5 - \sum_{s=3}^{Bt-1} sp(s) \qquad (5)$$

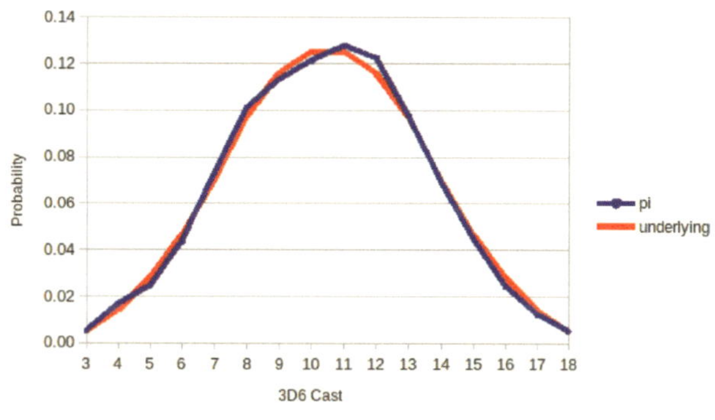

Figure 8.
Probability for 3D6 casts calculated from π made from the first 100,000 digits, converted to 20,000 dice casts and 6679 3D6 casts. These are compared to the underlying counted distribution for the 216 combinations of 3D6 given in ***Tables 1*** *and 2. This passes the Chi squared test [30].*

Bt	Calculated scores from Eq. (7)	Simulated scores Mean (Std dev)
3	10.50	10.53 (2.93)
4	10.55	10.60 (2.99)
5	10.69	10.65 (2.88)
6	10.97	11.00 (2.98)
7	11.41	11.45 (3.10)
8	11.99	11.96 (3.61)
9	12.64	12.48 (4.35)
10	13.14	13.05 (5.55)
11	13.33	13.34 (6.58)
12	13.10	13.13 (8.08)
13	12.50	12.68 (9.03)
14	11.68	11.76 (9.96)
15	10.90	11.06 (10.38)
16	10.27	10.41 (10.58)
17	9.84	9.90 (10.81)
18	9.61	9.49 (10.86)

Table 5.
Calculated scores for 3 $\leq Bt \leq 18$ using Eq. (7), compared to the results from 2160 simulations to 2dp.

where 10.5 is the average score for 3D6.

If you score less than Bt, then you roll again. If you score less than or equal to s, you score nothing. If you exceed that amount, you score:

$$Double = \sum_{s=3}^{Bt-1} \sum_{t>s}^{18} p(s)p(t)(s+t) \tag{6}$$

Hence the total average roll is:

$$Score = 10.5 - \sum_{s=3}^{Bt-1} sp(s) + \sum_{s=3}^{Bt-1} \sum_{t>s}^{18} p(s)p(t)(s+t) \tag{7}$$

For example, Suppose we choose Bt=4, then:

$$Score = 10.5 - 3p(3) + p(3) \sum_{t>3}^{18} p(t)(3+t) \tag{8}$$

This is:

$$Score = 10.5 - 3p(3) + p(3)[7p(4) + 8p(5) + 9p(6) + \ldots 21p(18)] \tag{9}$$

So using the probabilities identified in **Table 1** for each cast the average score per round is:

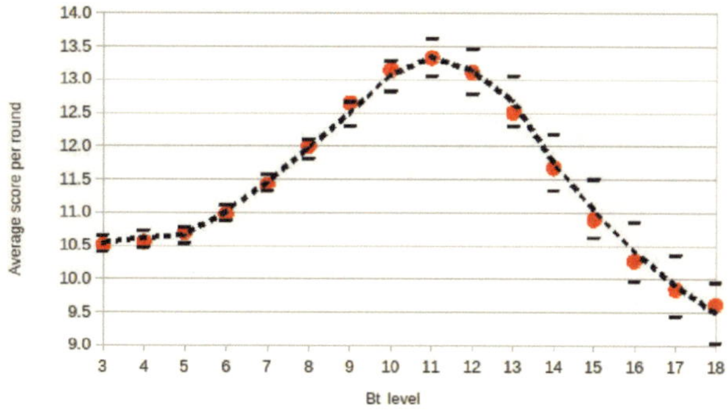

Figure 9.
Calculated average score (red) together with the mean from the simulated average scores for 3 ≤Bt≤18 from 2160 simulations linked by the dotted line and confidence interval markers at ± 95% levels. Data taken from ***Table 5****.*

$$Score = 10.5 - \frac{3*1}{216} + \frac{1}{216^2}[7*3 + 8*6 + 9*10 + 10*15 + 11*21 + 12*25 + 13*27 + 14*27 + 15*25 + 16*21 + 17*15 + 18*10 + 19*6 + 20*3 + 21*1] \quad (10)$$

$Score = 10.55$ for Bt = 4 to 2 dp's.

Following the same process, **Table 5** shows the calculation of the average score per round for $3 \leq Bt \leq 18$, together with the result of 2160 simulations performed by Google Sheets using the same rng as Roll Dice [6]. This uses the conversion from a random number to dice roll as discussed in Appendix 2. The data are plotted in **Figure 9** together with the confidence intervals calculated from Eq. (4).

The standard deviation increases as Bt increases due to the presence of more zero scores. Clearly an optimal average score per round exists at Bt = 11, and **Figure 9** shows this. Note that the calculated average score lies within the confidence intervals, indicating that the analysis is sound.

Author details

Mark Flanagan[1], Trevor C. Lipscombe[2], Adrian Northey[3] and Ian M. Robinson[3*]

1 NHS Business Service Authority, Newcastle upon Tyne, UK

2 Catholic University of America, Washington, D.C., USA

3 Hearts of Oak, North Yorks, UK

*Address all correspondence to: a.anser4u@gmail.com

References

[1] Strode T. A Short Treatise of the Combinations, Elections, Permutations & Composition of Quantities Illustrated by Several Examples, with a New Speculation of the Differences of the Powers of Numbers. Vol. 18. Godbid for Enoch Wyer; 1678. p. 19

[2] McShane JM, Ratliff MI. Dice distributions using combinatorics, recursion, and generating functions. The College Mathematics Journal. 2003;**34**: 370-376

[3] Anderson TW, Darling DA. A test of goodness-of-fit. Journal of the American Statistical Association. 1954;**49**:765-769. DOI: 10.2307/2281537

[4] The Anderson-Darling Statistic – Minitab Support. Available from: https://tinyurl.com/3f7mx9xw

[5] Balanda KP, Macgillivray HL. Kurtosis: A critical review. The American Statistician. 1988;**42**:111-119

[6] Roll Dice (Google App). Click Twice on the Six Dice to Generate 3D6, then Click Roll which Adds the Total for the Throw to Give the Cast. Available from: https://g.co/kgs/VYws2A

[7] Haramoto H, Matumoto M, Saito M. Unveiling patterns in xorshift128+ pseudorandom number generators. Journal of Computational and Applied Mathematics. 2022;**402**:113791

[8] Matsumoto M, Nishimura T. Mersenne twister: A 623-dimensionally equidistributed uniform pseudo-random number generator. ACM Transactions on Modeling and Computer Simulation. 1998;**8**:3-30

[9] Dodge Y. A natural random number generator. International Statistical Review / Revue Internationale de Statistique. 1996;**64**:329-344

[10] Jaditz T. Are the digits of π an independent and identically distributed sequence? The American Statistician. 2000;**54**:12-16

[11] Marsaglia G.. On the Randomness of Pi and Other Decimal Expansions. 2010. Available from: http://citeseerx.ist.psu.edu/viewdoc/summary?doi=10.1.1.694.4783

[12] Johnson BR, Leeming DJ. A study of the digits of π, e and certain other irrational numbers, Sankhyā. The Indian Journal of Statistics. 1990;**52**: 183-189

[13] Pathria RK. A statistical study of randomness among the first 10,000 digits of π. Mathematics of Computation. 1962;**16**(78):188-197

[14] pi to 100,000 Decimals Broken into Groups of 5 Digits. Available from: http://www.eveandersson.com/pi/digits/pi-digits?n_decimals_to_display=100000&breakpoint=5

[15] Litwiller BH, Duncan DR. The probability of winning dice games. The Mathematics Teacher. 1979;**72**:458-461

[16] Roters M. Optimal stopping in a dice game. Journal of Applied Probability. 1998;**35**:229-235

[17] Haigh J, Roters M. Optimal strategy in a dice game. Journal of Applied Probability. 2000;**37**:1110-1116

[18] Hill TP. Knowing when to stop: How to gamble if you must—the mathematics of optimal stopping. American Scientist. 2009;**97**:126-133

[19] Johnson RW. Toss up strategies. The Mathematical Gazette. 2012;**96**:415-419

[20] Crocce F, Mordecki E. A finite exact algorithm to solve a dice game. Journal of Applied Probability. 2016;**53**:91-105

[21] Liu L, Rettenmaier AJ, Saving TR. Staying the course or rolling the dice. Journal of Insurance Issues. 2019;**42**: 66-85

[22] Watson JM, Moritz JB. Fairness of dice: 'A longitudinal study of students' beliefs and strategies for making judgments. Journal for Research in Mathematics Education'. 2003;**34**: 270-304

[23] Nilsson P. Different ways in which students handle chance encounters in the explorative setting of a dice game. Educational Studies in Mathematics. 2007;**66**:293-315

[24] Rabin M. Inference by believers in the law of small numbers. The Quarterly Journal of Economics. 2002; **117**:775-816

[25] Silverman MP, Strange W, Silverman CR, Lipscombe TC. On the run: 'Unexpected outcomes of random events. The Physics Teacher. 1999;**37**:218

[26] Croson R, Sundai J. The Gambler's Fallacy and the Hot Hand: Empirical Data from Casinos. Journal of Risk and Uncertainty. 2005;**30**:195-209

[27] t-Testing and Significance. Available from: https://tinyurl.com/49c9u98h

[28] Holland JH. Genetic algorithms. Scientific American. 1992;**267**:66-73

[29] π to 100,000 Digits Converted to 20,000 Random Numbers and Dice Rolls. Excel file (.xls). Available from: https://tinyurl.com/2t8rtxz7

[30] Chi Squared Test Applied to Dice Rolls. Available from: https://tinyurl.com/4akxc9vs

[31] Diehard Tests. Available from: https://en.wikipedia.org/wiki/Diehard_tests

Chapter 3

Developing Inclusive Games: Design Frameworks for Accessibility and Diversity

Matheus Cezarotto, Pamela Martinez and Barbara Chamberlin

Abstract

All players should have the opportunity to play, engage with and enjoy a game, especially games that are designed to educate or transform the player. In addition to the game interface, mechanics and artwork, high quality games must also ensure that all players can use the controls, understand the context, receive information from the game, and have a sense of belonging to the world of the game, or ability to identify with messages and in-game worlds. Design teams may well have the desire to create games that are both accessible and representative of all players, but find the process of doing so overwhelming. Based on evolving work in an educational media studio, this chapter presents two frameworks to guide teams in reviewing the issues involved in accessibility, equity, diversity, inclusion and representation, with recommendations on steps to take towards integrating these design approaches into consistent development processes.

Keywords: inclusion, learning, process, a11y, DEI, representation, needs

1. Introduction

The ultimate goal of a learning game is to provide an engaging and effective learning experience, fostering a change to the player's knowledge, skill, behavior, emotion or physiology [1]. Extensive publications have offered guidance on making quality games, highlighting key markers of quality, identifying affordances of games, and justifying the potential of games to transform. Yet, fewer publications emphasize the importance of extending this access to all players, or ensuring that all players can see their culture, identity, and representation in the games they play. Inclusion is a state of being included in a group or activity. In game design, inclusion is the state of giving all players a sense of belonging to the game, messages or gameplay, as well as giving all players access to play, navigate and enjoy the full experience of the game. "The people who design the touchpoints of our society determine who can participate and who is left out... Design shapes our ability to access, participate in, and shape our world" ([2], p. 6–7). Inclusive design is often considered a bridging terminology to enable multiple types of backgrounds and abilities: including age, culture, economics, education, race and accessibility. "The focus is on fulfilling as many user needs as possible, not just as many users as possible" [3]. Because designers hold the power in making games inclusive, it may be helpful to view inclusivity in terms of two aspects: accessibility and diversity.

All players, regardless of their abilities, need a degree of accessibility to properly play and interact with a game [4]. Accessibility represents the features that developers design into the game to allow access and use of the game by players with a wide range of needs [4–7]. Accessibility is a large part of making games inclusive from a mechanical standpoint, in that accessibility usually gives a player access to ways of interacting with a game. Inclusivity also includes the way players see themselves in a game, identify with the content or intent of the game, and are given access to the game content in a meaningful way. This can include the presentation of characters, storylines and game environments that represent multiple worldviews, but issues of diversity, equity and inclusion also include fostering diverse teams, research, games, mechanics, and communities [8]. As more game developers prioritize accessibility, diversity, equity and inclusion in design, games will have wider appeal to more players, empowering and enriching players who are often underrepresented [9].

Researchers and developers have been making improvements in designing for accessibility, prompted in part by legislation such as, in the US, the Rehabilitation Act of 1973, and the Americans with Disabilities Act of 1990 [10]. The publication of the Twenty-First Century Communications and Video Accessibility Act [11] guided web development, which in turn prompted calls for greater accessibility in web-based games. Designing for accessibility has included several different terms, such as "Design for All" [12], "Barrier-free Design" [10], "Inclusive Design" [13], and "Universal Design" [10, 12]. Accessible design emphasizes the creation of products, systems and interfaces which includes all users.

Many individuals still face barriers in their daily life because their needs are not considered when products or systems are designed—which includes games and digital media. Prejudice and disablism against those individuals are still present in our society, disabling them from having equal opportunities to engage and participate in society [13]. In light of the "emancipatory research paradigm" [14] the authors acknowledge that users with disability are the true knowers of their challenges, desires, and needs; disabilities should not be falsely simplified as a tragedy to be pitied; researchers and developers need to foster an environment where users with disability can benefit by empowering them through participation; the research role should be in identifying and striving towards removing disabling social and physical barriers.

Research and development recommendations specific to diversity and equity are somewhat newer in game design, yet other media offers an insight regarding the need for representation. In a review of more than 160 studies on representation in media, Armstrong [15] documents a significant lack of representation of people of color, emphasizing the predominance of White characters in software, books and learning tools. Where progress has been made among representation of different racial, ethnic, and gender groups, characters who represent marginalized groups are still underrepresented. "Even when characters of diverse racial, ethnic, and gender groups are represented, these may not be accurate depictions. Some cases may reinforce stereotypes, limit portrayals and roles, and present inaccuracies" ([15], p. 30). To improve the situation, she calls for an improved sense of belonging in media for all individuals, cultural authenticity by engaging diversity among content creators, and recognizing nuanced identity by providing details of characters which reflect complex stories, relationships and situations [15].

The International Game Developers Association provides an overview of designing and developing inclusive games, with great emphasis on diversity. The framework in "Inclusive Game Design and Development" offers suggestions for strengthening overall development, starting with a diverse team, thorough research, world building, characters, design and mechanics, and even marketing [9].

2. Designing for inclusion: challenges and benefits

Facilitating dialog that leads to change can be challenging within design and development studios. Creating effective learning games is a complex task; incorporating inclusive design approaches can add extra challenges for design teams. An inclusive game design provides access to the game but also can foster acceptance and belonging among players [16, 17]. However, engaging in discussions about inclusion identifies gaps where others are still excluded. In a broad picture, both **accessibility** and **diversity** involve acknowledging the existence of people and/or groups who lack social power, prestige, or entrenched advantage, as well as establishing an awareness of their marginalization or exclusion based on social or physical barriers [16].

Developers of learning games experience additional challenges to making their games accessible. In addition to issues related to user interface and communication of game status to the player, games which seek to change the player in a meaningful way must also balance the learning goals and content-specific educational approaches with the needs of users. For example, mathematics is traditionally taught with visual guides such as maps and graphs: translating facility with these mechanics into gameplay can be challenging. Adding full accessibility to these games presents an additional level of design considerations.

By making games accessible and addressing diversity, equity and inclusion, development teams not only benefit players but also make their games better. Blackwell [18] coined the term "the Curb Cut Effect", to denote instances when accessibility changes benefited unintended audiences: curb cuts in sidewalks help people in wheelchairs move from the paved sidewalk to cross a street, as well as parents with strollers and delivery drivers with carts. A game that provides subtitles for the narrative offers benefits to users who are deaf as well as to those who are playing in a second language, those who cannot use sound at the moment, and those who need more time to cognitively process the information. Designing games towards accessibility also increases the game value as a meaningful experience for players and the community, fostering equality and a relevant social connection between players with and without disabilities [19].

Because curb cuts are now commonplace, their origins in accessibility are sometimes forgotten, without recognizing them as having been intentionally designed. Similarly, designing for diversity must also be intentional in providing access to multiple stories, identities and cultures. It gives learners—who may not traditionally have the opportunity to see themselves in materials—a mirror through which they see people like them, and gives all learners proficiency in a "broader range of expressive, analytic, and cognitive styles that are crucial to success in the twenty-first century" ([20], p. 17).

Progress is being made in both fields. One industry leader, Unity, offers an online free practical course [21] on game accessibility, supporting teams to design games for diverse audiences, with modules considering the design and development process. Microsoft has been very active in supporting the development of accessible media, with special attention to the gaming community. From their initiative it is worth highlighting the online course "Gaming accessibility fundamentals" [22] and the inclusive design toolkit [23]. Schell games provides an online accessibility matrix [24] that allows developers to identify, discuss, prioritize and find solutions to accessibility needs. A group of studios, specialists, and academics organized and keep updating the "Game Accessibility Guidelines," [25] a reference list for designing inclusive games. Able Gamers Charity [26] has been active in making an impact in players with disabilities' lives by providing peer counseling, engineering research on assistive technology, community development, user research to find solutions to barriers, and professional development promoting inclusion.

In our evolving world of social justice, more developers, design teams, and educational institutions are acknowledging the importance of **diversity, equity and inclusion (DEI)**, a trending topic. There are many published works regarding the wider scope of the topic. Rittner [27] provides a critical review of systems such as the *institutional*, *economic*, *social*, *political* and *interpersonal* as a way to recognize the need for change and how designers can create impact through their work. Alozie et al. [16] provide an overview of the importance of designing for equity and inclusion and the impact on student outcomes when consideration for the diverse needs of diverse students is taken into account.

In dividing *inclusion* into two categories of *accessibility* and *diversity*, this chapter attempts to guide design teams by offering two simple frameworks for understanding key components, facilitate discussion among developers, and refine processes used in development to address each aspect in accessibility and representation. While each concept is complex, teams can begin by identifying the specific areas that must be addressed, then translating them into steps relevant to their own design processes.

3. Frameworks for inclusion: accessibility and diversity

Several barriers may keep teams from engaging in the design of inclusive games, which are consistent with fears articulated by Holmes about inclusion in general [2]. Individuals may be fearful about engaging in the discussion, out of fear of using the wrong vocabulary or unintentionally hurting others; designers may fear that they cannot solve the entire problems, so it is not worth trying; fear of scarcity means worrying there aren't enough resources, such as people, money and time.

Simplified frameworks can help teams familiarize themselves with the key issues, build their vocabulary around inclusion, and begin to identify reasonable steps they can take to address parts of each problem. Inclusion as a broad goal suggests games are available to all players, but design teams may find it easier to break inclusion into two types of actions: accessibility and diversity. Simply: frameworks are a gateway for each design team to establish best practices for inclusive design of the games they create.

3.1 Framework for designing games for accessibility

Accessibility in games represents a set of characteristics that developers design into a game to provide players access, considering players needs related to vision, hearing, motor skills and cognition [28]. By understanding game accessibility from the *social* model of disability, in contrast to the *medical* model, accessibility lives in the product and not in the user [29]. It means that the disability is a mismatch between the design and the person's needs, instead of a personal health condition. Thus, a good game design that matches users' needs enables players, and a bad game design that does not match users' needs disables players.

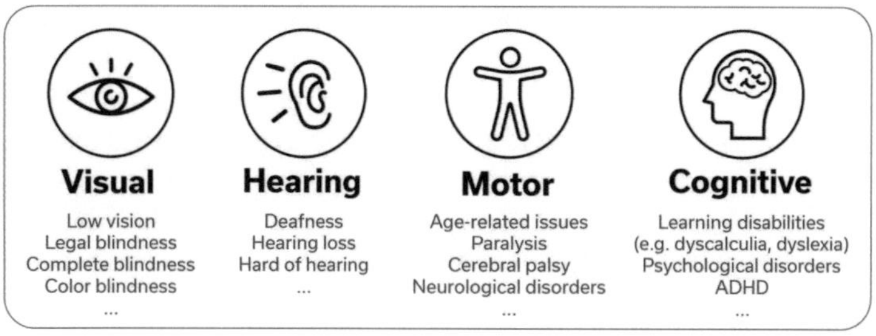

Figure 1.
Framework for designing games for accessibility.

All players fall somewhere on a spectrum of need in three possible scenarios: permanent need (e.g., a person with one hand), temporary need (e.g., person with a hand injury), and situational need (e.g., a new parent holding a baby) [23]. Identifying and addressing those needs can be a challenging task, since there is a wide range of needs. One approach to support this task is to discuss the needs in categories, being able to see an overview picture. In the literature four main categories of needs are used—**visual**, **hearing**, **motor**, and **cognitive** [30]. These categories allow the identification of possible barriers players may face in games (**Figure 1**).

- **Visual needs:** players with some degree of vision loss, such as low vision, legal blindness, complete blindness, color blindness.

- **Hearing:** players with some degree of loss in the ability to hear, either from one or both ears, such as hearing loss, hard hearing, deafness.

- **Motor:** players with some mobility limitation or muscle control, such as cerebral palsy, lack of steadiness, lack of mobility, age-related issues, neurological disorders, repetitive stress injury, paralysis, arthritis.

- **Cognitive:** players with mental or psychological disorders, which causes a deficit in the ability to learn, process or remember information, communicate, make social interaction, and make social decisions. Examples include neurodivergent learners, learning disabilities, or intellectual disability.

Each area of need exists within a spectrum, from low to high, and there are specific types and profiles within each category—for example, two players with ADHD (attention-deficit/hyperactivity disorder) can have different cognitive needs to support attention. Instead of having discrete boxes for needs and disabilities, they are frequently co-diagnosed, with any given user having needs across several different types of issues—for example, a deaf player will likely have needs associated with hearing, and may also have needs associated with some cognitive functions.

3.2 Framework for designing games for diversity

Diversity in games represents components of inclusion that developers incorporate into their teams and games for broader accessibility. Diversity, equity, inclusion (DEI), and accessibility have come to the forefront of education and societal issues, forcing developers to focus on addressing them in more meaningful ways to support social imbalance. As specializing in DEI continues to evolve, so do references to it. Often seen as DEI in the past, it is evolving to represent the order of concentration, Equity, Inclusion, and Diversity (EID). Diversity in relation to games is observable in how diverse a development team is (who is part of the team, who is making decisions, who has the power) and transparency in knowledge and sentiment towards inclusion. Moreover, it includes how this design team portrays game worlds, environments, storylines, and characters in their games to promote identification, acceptance, and counteract stereotypes.

In creating a framework, the authors drew from the expertise of the International Game Developers Association Diversity white paper that provides a unique structure. This paper offers a more succinct framework for diversity with a more in-depth perspective on accessibility to strengthen inclusion practices in game development (**Figure 2**).

- **Team Building:** Design and development teams include professional instructional designers, educators, learning specialists, animators, web developers, programmers, and marketing and distribution experts with various educational backgrounds and expertise. In addition to the professional roles, each team likely includes individuals of different ethnicities, gender identities, varied backgrounds, and personal experiences. This richness contributes to the creativity needed to produce inclusive educational media, and should be encouraged in hiring practices: diverse teams are one of the easiest ways to make sure diverse voices are heard. Regardless of the makeup of any given team, professional development opportunities for diversity, equity, and inclusion coaching can increase the ability of a team to meet inclusion needs. Meet annually to discuss issues, engage in guided case study reviews, exchange ideas, and examine fresh viewpoints and perspectives. In addition, it takes intentional actions to curate a culture of inclusion where everyone has a voice, and to reflect on power dynamics. Even when it is assumed that all voices are welcomed, unseen power dynamics may make some individuals feel as though their voices are not encouraged. Ultimately the quality of the products represents the dedication of focus on accessibility and diversity for a broader range of inclusivity for players. This work should be ongoing, consistent, and reflective throughout the life of a studio.

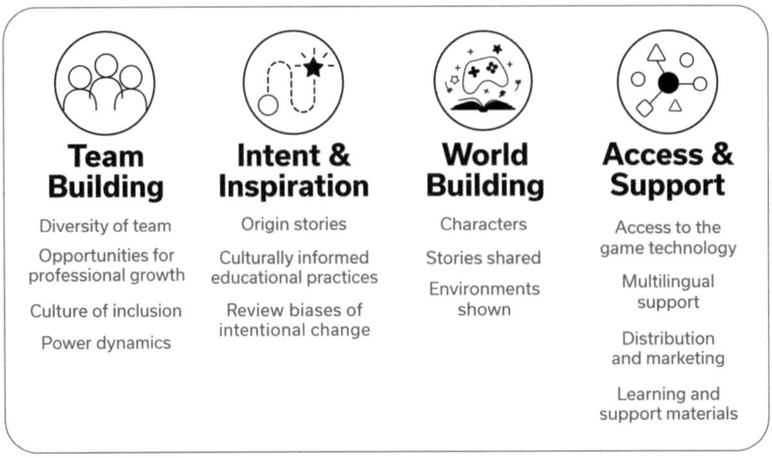

Figure 2.
Framework for designing games for diversity.

- **Intent, and Inspiration:** When the idea for a game first appears, it may be rooted in an educational goal, a specific audience to address, or simply a great idea. These origins are often rooted in a given perspective or experience of an individual or group of individuals. When a development team is culturally informed, they research the target audience's social, cultural, and ethnic dimensions; evaluate their own biases in making assumptions about needs; and reflect on the origin of stories, themes and visual presentation. Through these efforts, developers can design appropriate origin stories, themes, characters, and storylines inspired by situations around the topic and content. Continuous discussions help to address biases and support intentional change. Design and development teams should include content experts or advisors to align content and educational support to determine appropriate technologies for target audiences.

- **World Building:** In-game experiences are another way to support inclusion through diversity: what does a player see and hear while playing? How are the characters, locations and storylines presented? Discussion around the environment, geography, and characters happens early in the design of the game and should include discussions centered on diversity. Development teams can spend months planning and defining game worlds and esthetics. The mapping, environment designs, geography, and characters must support cohesive designs towards inclusivity to engage players fully. Consideration of themes, cultures, and people should happen during this development phase through extensive research; representation is at the forefront during this part of the process. A diverse team allows for exceptional discussion of shared experiences and stories that strengthen design decisions. Consult content experts, accessibility and diversity specialists. Not all games may follow realistic cultures and societies, and it is important to review similarities in fictional worlds to real worlds. Designers and developers want to be mindful of crossovers to avoid common tropes or harmful stereotypes. This awareness fosters acceptance and belonging, allowing for a broader range of inclusion.

- **Access and Support:** Games developed in educational development studios may be created to specifically address the needs of marginalized groups, but all games should be created in ways that support the use of all players. Who will access the game, and how does access to technology create barriers for some players? Research helps to determine the type of technology available and the medium to pursue distribution, such as on mobile, tablets, or web, as well as connectivity issues. Addressing technological and educational needs to address equity is a vital conversation to have before development begins, especially when working with marginalized communities. What role does language play in access? Research the target audiences' educational needs; investigate the possibility of multilingual adaptations and support materials. How do cost or subscription requirements limit some audiences? Distribution and marketing efforts should support the target audiences but may also require larger discussions about reaching wider audiences.

4. Process considerations

The Learning Games Lab is a university-based, non-profit development studio that develops educational games, virtual labs, videos, animations and other interactive tools to help learners of all ages learn from research-informed interventions. As part of NMSU's Innovative Media Research and Extension department, the studio partners with research groups, faculty, and programs nationally and internationally to create educational media in various disciplines. Design-based research guides Learning Games Lab product development, informing the specific goals of each learning game, the audience, as well as the final product.

The Learning Games Lab has used both frameworks in developing their own processes regarding accessibility and representation. Their work occurs throughout the life of the studio, as well as at specific points in the development of a single game. All team members share responsibility in designing for inclusion, and their processes are evaluated continuously through their interactive design process. The team also discusses and integrates accessibility features into products beginning with early designs, so a wide range of users can use the products without interaction barriers.

Action research enables the teams to actively work to create change in their process to prioritize accessibility and diversity, following a continuous, ongoing, and iterative process, a self-reflective spiral cycle of *planning*, *acting*, *observing*, and *reflecting* [31, 32]. Additionally, the team investigates the work of other developers, community organizations, and experts to inform their own development practices, and revisits previously developed games to improve their accessibility and inclusion.

In using the frameworks to understand the key components of inclusion, the team works through specific steps to address the needs in every product created. This includes five specific processes (**Figure 3**).

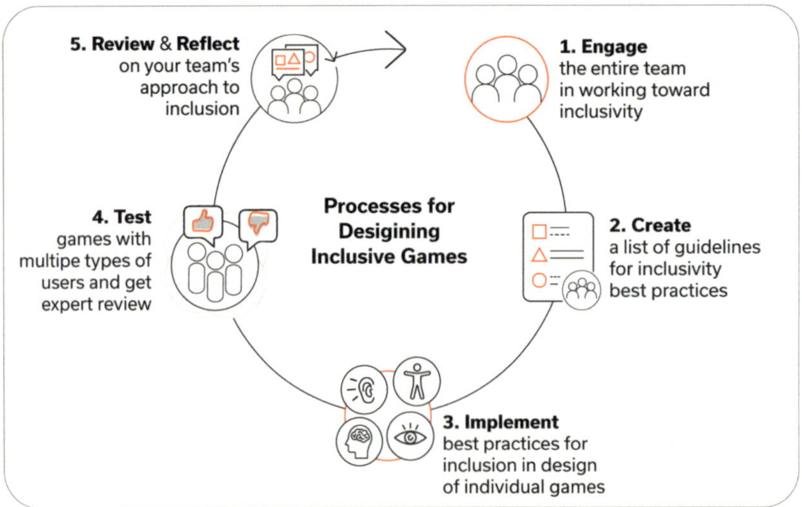

Figure 3.
Processes for designing inclusive games.

4.1 Recommendations for design teams

1. **Engage the entire team in working towards inclusivity:** Create and foster an accessibility and diversity, equity and inclusion mindset among your team. The entire team should understand the value of inclusive games, have vocabulary and be given opportunities to talk about it. Establishing the core principles for your studio can establish lists of what accessibility and inclusion mean, and may provide valuable examples for reference in development.

For example, the Learning Games Lab design team has defined their core principles to include [33]:

- **Representation**: to include a variety of positive visual, cultural and historical and economic presentations so that users can see themselves and others in the products they use. We have articulated our specific approach to equitable representation, and re-visit and edit this approach annually.

- **Accessibility**: enabling use and enjoyment of our products for individuals on a spectrum of physiological needs, such as vision, hearing, motor control and cognition. We also seek to provide diversity for various types of access based on economics, location, and technology. We *use a* specific framework and approach for thinking through accessibility needs, and update this approach annually.

In developing the core principles, it is helpful to engage in other opportunities.

- **Review old games:** Reviewing old projects is a good opportunity for the team to identify accessibility barriers, reflect about things learned and discuss ways to make the product more accessible. Assign at least two members of the team to make an inclusivity review on an old game. Ask reviewers to present the accessibility issues identified. As group, discuss, prioritize and find solutions to remove the accessibility barriers. Similarly, ask reviewers to challenge the assumptions held in the game about the intent and cultural biases of the game, and to identify any instances of cultural appropriation, tropes, or misrepresentation. Conduct a character or world review to track who is and is not represented, as well as the types of environments and stories told. Reviewing old projects gives the team a safe space to normalize and foster a vocabulary on inclusion; additionally, the team gets smarter for future projects.

- **Support professional development:** Encourage and support your team to find knowledge outside of the studio. Interaction with other studios is a good way to learn how other developers address similar issues. Participating in conferences and workshops also give team members an opportunity to learn and connect with other developers and researchers. Present or talk about your team experience with accessibility in conferences; this brings a chance for connection and also allows team members to reflect on the situation, considering different perspectives. Increase participation and expertise through professional learning networks (PLN) that support diversity, equity, inclusion and accessibility initiatives such as educational coaching, webinars and PLNs on social media.

- **Bring specialists to talk with your team:** Bring inclusion specialists to review your games and talk with your team. Your team will learn about possible interaction barriers in the game, and also will better understand the accessibility review process and its value. Invite diversity, equity and inclusion specialists into conversations early and ask them to review throughout the development process.

- **Work and revisit accessibility and representation statements:** As a team, develop core value statements regarding your approach to inclusion. Your team should review and evaluate the statement at least once a year. Give an opportunity to the whole team to participate in creating and refining the statement.

Create a living document that defines a set of guidelines or principles for diversity, equity, inclusion, and accessibility for the team to follow. Review and refine it yearly as the team becomes aware of new information and research that could benefit design and development teams.

- **Learn from accessible and diverse games:** Accessibility for game design and diversity is gaining attention; studios and researchers are increasingly sharing approaches and best practices for accessibility and representation in games. The release of games (from Indie to triple AAA) with accessibility features, and with a diverse cast of characters and stories, is escalating. Play those games and learn from their approaches. As developers we get better by playing and reviewing games. Give opportunities to your team to play inclusive games, and create a space where they can share what they learn with the team.

2. **Create a list of guidelines for inclusivity best practices:** Create a document with accessibility and representation guidelines for all game projects; arrange that the team can easily access and update this document. Every game project will use the guidelines differently. Thus, your guideline list should be seen as a guiding point, which the team will use to discuss, reflect and creatively make inclusive design decisions. Revise your team's guidelines based on technological advancements, learning from projects, and interaction with other studios or experts. Your team can create their own guiding principles for inclusion using different approaches and formats, including:

- **Select from existing guidelines:** There are several excellent and detailed guidelines available—some of them were mentioned earlier in this chapter. From those, select and create your team guideline document. This can help avoid the feeling of being overwhelmed with so much information that some things are not applicable to the game your team is working on, or to the scope and budget available for the project.

- **Use the frameworks:** The four main areas of need for accessibility—*Vision*, *Hearing*, *Motor* and *Cognitive*—is an effective way to organize guidelines and game accessibility features. The areas of need give the team an overview picture, support the identification of barriers or specific issues within the game and game team, and also give the team a vocabulary so they can talk about and articulate accessibility decisions. The four main areas of diversity in games—*Team building*, *Intent & Inspiration*, *World Building*, and *Access and Support*—help teams to discuss and design for diversity. Consider posting these frameworks for the team to access in generating discussion

- **Determine levels of priority for the project:** Inclusion is complex and can be overwhelming, and it can present contracting indications. For example, in designing one character to specifically include one group, it eliminates the possibility of another group being represented in that group. Similarly, creating a captioning process to meet audio needs may present visual challenges. By prioritizing the design and implementation of certain accessibility features based on the project scope, budget and intent, the team is better able to plan and make smart design decisions specific to the needs of a given project.

- **Foster inclusivity conversations among your team:** Allow time for teams to meet and have those difficult conversations that may come with diversity, equity, and inclusion. If necessary, bring in a specialist to guide the conversation; it could save time overall. Meet annually to discuss the guiding principles and approach to Diversity, Equity, Inclusion, and Accessibility. Make sure reviewers—in-house, invited, or hired—are composed of diverse team members. Remember, everyone brings different experiences and voices to the table.

In creating their list of guidelines, the Learning Games Lab defines representation and accessibility actions in the following ways [32]:

4.2 Representation: how the Learning Games Lab portrays individuals in the media it creates

In our products, we seek to produce media that offers diverse representation, promotes acceptance, includes all learners, and counteracts stereotypes. Individual differences and attributes may include:

- **gender** (acknowledging the spectrum);
- **sexual orientation**;
- **body shape** (weight, height, development, anthropometrics);
- **voice,** accent, dialect, way of speaking, vocabulary;
- **skin**, **eye color** and **hair color** and style;
- **clothing and vestments**, including culturally specific or faith-based;
- **age**;
- **social class;**
- national **origin**, **location** (city, suburban, rural);
- **social relationships** (such as living arrangements or family structure);
- **abilities** and **preferences** (such as physical, cognitive, motor or social); and
- **beliefs** (such as religious or atheist).

4.3 Accessibility: how the Learning Games Lab helps people use and enjoy its products

Making our products accessible is a **primary design consideration** for our studio. We strive to make our products as accessible as possible, knowing that we will have to make some compromises, and may fall short in some areas. Our efforts recognize that:

- Accessibility is not just for a set group of users: **all users fall somewhere on a continuum of need** (permanent, temporary, situational). Accessibility is about user needs, and accessible features must be actively designed during the design process.

- **Disabilities do not exist in discrete boxes:** they are often co-diagnosed, with any given user having needs across several different types of issues. Each area of need exists within a spectrum, from low to high, or specific types within each category.

- **Accessibility lives in the product and not in the user:** a bad design that does not match users' needs disables people, and a good design that matches users' needs enables people.

- User **variability** may include:

  - **Visual needs**: the person has a certain degree of vision loss, such as low vision, legal blindness, complete blindness, color blindness. This means our products should be reviewed for contrast, color and on-screen text or visual cues — providing alternatives for users.

  - **Hearing needs**: the person has a certain degree of loss in the ability to hear, either from one or both ears, such as deafness, hearing loss, or hard hearing. This is met by offering captioning of both spoken text and other audible cues.

  - **Cognitive needs**: the person has a mental or psychological disorder, which causes a deficit in the ability to learn, process or remember information, communicate, make social interaction, and make decisions. This type of disability can be a learning disability, intellectual disability, or a specific cognitive ability (e.g., memory, language processing). Includes developmental disabilities (e.g., dyslexia, dyscalculia), attention deficit hyperactivity disorder (ADHD), Alzheimer or senility because of aging, people with autism, Down syndrome, and other mental retardation. Some people with cognitive issues need information in literal language comprehension. Their thinking is more concrete, rather than abstract. We address this in a wide variety of ways, including design which offers explicit cues and expectations to guide users with cognitive needs.

  - **Motor needs**: the person has a limitation or a loss in the mobility function and muscle control, such as arthritis, paralysis, repetitive stress injury, neurological disorders, age related issues, lack of mobility, lack of steadiness, or cerebral palsy. We address this by developing resources with interfaces which can easily be mapped to alternative controllers.

3. **Implement best practices for inclusion in design of individual games:** Design your game to be played by as many players as possible, considering accessibility needs (vision, hearing, motor and cognitive) and diversity of identities. Work with your team to anticipate possible gameplay and interaction barriers based on user needs, and use this information to guide design decisions.

   - **Design for accessibility and diversity from the beginning:** Make accessibility and diversity discussions and decisions part of the entire design and development process. Instead of something extra, think with your team about ways to improve players' learning experience by improvising accessibility needs.

- **Follow best practices and your team's guidelines** to design inclusive experiences. Ideally, use extensive formative user testing during development to assess how the game supports user's needs.

4. **Test games with multiple types of users and get expert review:** User testing always yields the most direct feedback from players, particularly if you use multiple methods to get feedback, such as observational data, focus groups, play-aloud narration or individual interviews. However, one key aspect of inclusion is recognizing the specific and unique needs of any given player. This can create unintentional biases in user testing: for example, asking one Latina player her perspective on a main character cannot be interpreted as implying the same experience for all Latina players. Similarly, a player with a visual impairment may be able to successfully use one type of web browser to play, when another may not. Where possible, encourage a wide range of players to access your game, and encourage additional strategies to complement your testing.

- **Ask for gatekeeper and expert review:** What other experts could conduct a review of your game? If you have access to others who support a wide range of accessibility needs or educators who work across specific issues in representation, ask them to provide an analysis of your game. This second-level review may not be as personal as going to individual users, but it can provide considerations that warrant another review by others.

- **Consider paid quality assurance reviews**: Investigate organizations who may be able to provide and coordinate reviews for accessibility, testing across different devices or with different groups.

- **Encourage ongoing feedback:** In accepting that you cannot possibly get enough feedback from *all of your audiences*, enable feedback after the game is delivered or launched. Offer an email or other pathway for requests or suggestions for improvement. Consider a link that specifically mentions the game's commitment to inclusion, and invites users to respond with concerns or recommendations specific to your inclusion goals.

5. **Review and reflect on your team's approach to inclusion:** Give your team opportunities to reflect on your accessibility and representation initiatives, being critical about what is working and where some improvements are needed. There are several ways to approach inclusive design rather than a single solution that applies for all. With that in mind it is important that your team finds what works for you, considering your scope of work, budget, technical knowledge and so on. Your team can find the process that works for you in various ways, including:

- **Recognize that inclusive design is an ongoing process:** Keep in mind that inclusive design in learning games is complex, and it is an ongoing process. Work with your team to prioritize accessibility features that are feasible to be implemented in your games, and plan strategies to allow your team to keep adding more and more features on projects.

- **Create an accessibility and representation plan:** Create a plan for your team in terms of accessibility and representation. Your team can start with small and basic features to make standard in your games, for example, the use of appropriate fonts, text size, checking color for contrast and color blindness, and subtitles. Once you have some standard, start planning ways to get more accessibility features into your games, in terms of what your team needs to learn or what technical support is needed for that.

- **Reflect:** The Learning Games Lab has used action research to enable this self-reflection and identify implications for the studio design process. This approach was effective to support an accessibility mindset among the team, creating a vocabulary that allows the teams to discuss and make accessibility decisions.

- **Teach others about what your team has learned:** An effective way to reflect and improve your game accessibility initiatives is to teach what you have learned so

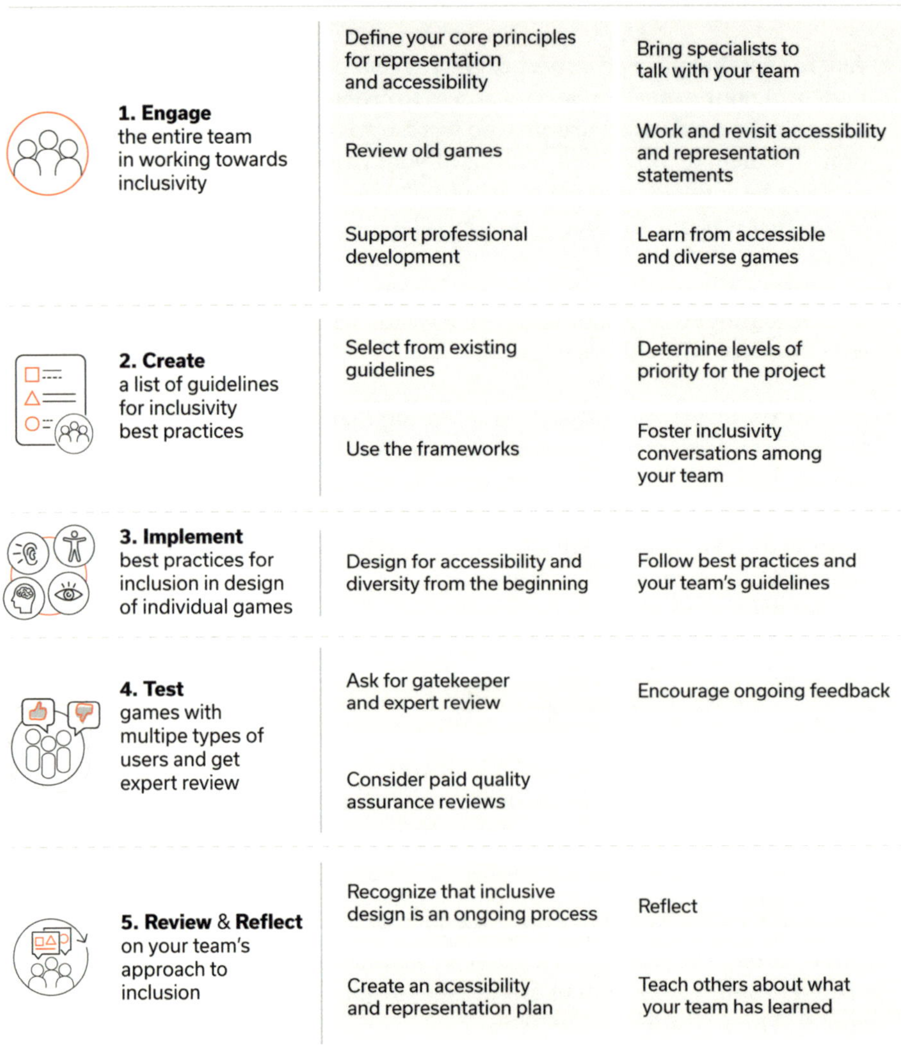

Figure 4.
Recommendations for design teams - flowchart.

far to others. Conference workshops or other teaching opportunities allow your team to reflect about your process and also receive feedback on it. In **Figure 4** we provide a visual summary of the recommendations outlined in this study.

5. Conclusions

Inclusivity in games represents a design quality which allows users with a wide range of skills, needs, and backgrounds to resonate with a game and have a great experience. All players need a degree of accessibility to properly play any given game. Players need mirrors to see themselves and windows to see what they could be and how others live [15]. When players' needs are not taken into consideration, interaction barriers occur in the game, preventing players from using the system or delivering them a bad experience. Inclusive games support players' visual, hearing, motor, and cognitive needs. In addition to enabling access, the game should be easy to use, provide a good user experience, allowing players to thrive through game challenges, and allow players to identify themselves with the game. Inclusive games are developed by diverse teams which emphasize the growth of the entire team and review the intent and inspiration of the games in development. Inclusive games offer worlds rich in perspectives, visual and cultural presentation, which consider access and support in distributing the game. Grounded in development and research experience, processes for designing for inclusion should happen throughout the life of a studio by engaging the entire team, developing core values and specific guidelines, then implementing those in an interactive cycle of designing, testing and modifying games. The review of completed projects, as well as ongoing practices, keeps a design team up to date in their initiatives. Finally, teams with successful approaches and recommendations can share their progress with the industry, to ensure that games developed in the field continually improve.

Acknowledgements

We wish to thank the developers in the Learning Games Lab at New Mexico State University. The team's constant efforts to improve accessibility in their games has enabled this publication: We recognize the contributions of everyone on our team in making games better and helping us articulate these processes for others. Special thanks to Amanda Armstrong for guidance, wording and continued work on issues related to diversity and representation. We also acknowledge the contributions of Bridge Multimedia (specifically, Matthew Kaplowitz) for consultation and advice, which contributed tremendously to the process initiated in this work, and Amy Smith Muise for improving the written quality of this paper.

Conflict of interest

The authors declare no conflict of interest.

Author details

Matheus Cezarotto*, Pamela Martinez and Barbara Chamberlin
Learning Games Lab, New Mexico State University, Las Cruces, USA

*Address all correspondence to: matheus@nmsu.edu

References

[1] Chamberlin BA, Schell J. Connected Learning Summit, The Secret Process for Making Games That Matter. Cambridge, MA: MIT Media Lab; 2018

[2] Holmes K. Mismatch: How Inclusion Shapes Design. United Kingdom: MIT Press; 2020

[3] Joyce A. Inclusive Design. Fremont, CA: Nielsen Norman Group; 2022 [cited September 16, 2022]. Available from: https://www.nngroup.com/articles/inclusive-design/

[4] Hersh MA, Leporini B. An overview of accessibility and usability of educational games. In: Student Usability in Educational Software and Games. Hershey, PA: IGI Global; 2013. pp. 1-40

[5] Preece J, Sharp H, Rogers Y. Interaction Design: Beyond Human-Computer Interaction. Chichester: Wiley; 2015

[6] Engelen J. Inclusive Design Guidelines for HCI. United Kingdom: CrC Press; 2001

[7] International Organization For Standardization. ISO/TS 20282-2: Usability of Consumer Products and Products for Public Use – Part 2: Summative Test Method. Geneva, Switzerland: ISO; 2013 [cited September 16, 2022]. Available from: https://www.iso.org/obp/ui/#iso:std:iso:ts:20282:-2:ed-2:v1:en

[8] Sharp D. IGDA Report Tackles the Essentials of Inclusive Game Design and Development. San Francisco, CA: VentureBeat; 2021 [cited September 16, 2022]. Available from: https://venturebeat.com/games/igda-report-tackles-the-essentials-of-inclusive-game-design-and-development/

[9] Gittins RK. Inclusive Game Design and Development White Paper. Toronto, ON: Facebook Gaming, International Game Developers Association; 2022 [cited September 16, 2022]. Available from: https://igda-website.s3.us-east-2.amazonaws.com/wp-content/uploads/2021/12/08124833/Inclusive-Game-Design-and-Development.pdf

[10] Keates S, Clarkson PJ, Harrison L-A, Robinson P. Towards a practical inclusive design approach. In: Proceedings on the 2000 Conference on Universal Usability - CUU '00. New York, New York, USA: ACM Press; 2000

[11] Federal Communications Commission. 21st Century Communications and Video Accessibility Act (CVAA) Consumer Guide. Washington, DC: The US Federal Communications Commission; 2010 [cited June 11, 2020]. Available from: https://www.govinfo.gov/content/pkg/PLAW-111publ260/pdf/PLAW-111publ260.pdf

[12] Vanderheiden G, Tobias J. Barriers, incentives and facilitators for adoption of universal design practices by consumer product manufacturers. In: Proceedings of the Human Factors and Ergonomics Society Annual Meeting. Vol. 42, No. 6. Newbury Park, CA: SAGE Publications; 1998. pp. 584-588. DOI: 10.1177/154193129804200613

[13] Gilbert RM. Inclusive Design for a Digital World: Designing with Accessibility in Mind. United States: Apress; 2019

[14] Stone E, Priestley M. Parasites, pawns and partners: Disability research and the role of non-disabled researchers. The British Journal of Sociology. 1996;**47**(4):699-716. DOI: 10.2307/591081

[15] Armstrong AL. The Representation of Social Groups in U.S. Educational Materials and Why it Matters: A Research Overview [Internet]. Washington, DC:

New America; 2021 [cited September 16, 2022]. Available from: http://files.eric.ed.gov/fulltext/ED619069.pdf

[16] Alozie N, Lundh P, Laguarda K, Parker CE, Fujii R, McBride B. Designing for Diversity Part 1. Where is Equity and Inclusion in Curriculum Design? [Internet]. Rockville, MD: National Comprehensive Center at Westat; 2021 [cited September 16, 2022]. Available from: https://compcenternetwork.org/sites/default/files/SRI_Paper%201_%20D4D_SRIrevisions_Final.pdf

[17] Williams DA, Berger JB, McClendon SA. Toward a Model of Inclusive Excellence and Change in Postsecondary Institutions. Washington, D.C.: Association for American Colleges & Universities; 2005 Available from: https:/perations.du.edu/sites/default/files/2020-04/model-of-inclusive-excellence.pdf

[18] Blackwell AG. The curb-cut effect. Stanford Social Innovation Review. 2017;**15**(1):28-33

[19] Cairns P, Power C, Barlet M, Haynes G, Kaufman C, Beeston J. Enabled players: The value of accessible digital games. Games Cult. 2021;**16**(2):262-282. DOI: 10.1177/1555412019893877

[20] Chita-Tegmark M et al. Using the universal design for learning framework to support culturally diverse learners. Journal of Education. 2012;**192**(1):17-22

[21] Unity Technologies. Practical Game Accessibility [Internet]. San Francisco, CA: Unity Learn; 2022 [cited September 16, 2022]. Available from: https://learn.unity.com/course/practical-game-accessibility?tab=overview

[22] Gaming Accessibility Fundamentals. Microsoft.com. [cited September 16, 2022]. Available from: https://docs.microsoft.com/en-us/learn/paths/gaming-accessibility-fundamentals/

[23] Microsoft design. Microsoft.com. [cited September 16, 2022]. Available from: https://www.microsoft.com/design/inclusive/

[24] Govender A. Schell Games-Making Experiences We're Proud of With People We Like to Make the World A Better Place. Innovations of The World. Innovationsoftheworld.com; 2021 [cited September 16, 2022]. Available from: https://innovationsoftheworld.com/schell-games-making-experiences-were-proud-of-with-people-we-like-to-make-the-world-a-better-place/

[25] Game Accessibility Guidelines: A Straightforward Reference for Inclusive Game Design. [cited September 16, 2022]. Available from: https://gameaccessibilityguidelines.com

[26] AbleGamers. Combating Social Isolation through Play [Internet]. Kearneysville, WV: The AbleGamers Charity; 2021 [cited September 16, 2022]. Available from: https://ablegamers.org

[27] Rittner J. Design education reform: Modeling equity and inclusion in teaching and learning. Design Management Review. 2020;**31**(3):12-22. DOI: 10.1111/drev.12236

[28] Cezarotto MA, Chamberlin B. Towards accessibility in educational games: A framework for the design team. Info. 2021;**18**(2):102-113. DOI: 10.51358/id.v18i2.931

[29] Oliver M. The social model of disability: Thirty years on. Disability & Society. 2013;**28**(7):1024-1026. DOI: 10.1080/09687599.2013.818773

[30] Cezarotto M, Martinez P, Chamberlin B. Redesigning for

accessibility: Design decisions and compromises in educational game design. International Journal of Serious Games. 2022;**9**(1):17-33. DOI: 10.17083/ijsg.v9i1.469

[31] Kemmis S, Mctaggart R. Participatory action research: Communicative action and the public sphere. In: Denzin N, Lincoln Y, editors. The Sage Handbook of Qualitative Research. California: Sage publications Inc; 2005

[32] Muratovski G. Research for Designers: A Guide to Methods and Practice. United Kingdom: SAGE Publications; 2016

[33] Innovative Media Research and Extension. Core Values of the Innovative Media Research and Extension Department. 2022 [cited September 16, 2022]. Available from: https://innovativemedia.nmsu.edu/about/principles.html

Chapter 4

Games and Social Reality

Yasuo Nakayama

Abstract

In this chapter, I will show that social activities can be seen as activities in games. In Section 2, rested on a formal framework called Dynamic Belief-Desire-Obligation Logic, I propose a model of agents who can play games. In Section 3, based on this framework, a rule system of a game is interpreted as a normative system that determines the obligation space and the permission space of each player. In a game stage, each player fulfills her obligations and chooses an action type from their permission space and performs it. These actions change the state of the game and determine new obligation and permission spaces of all participating players. In Section 4, I interpret social actions as actions based on normative systems. Social normative systems determine obligation and permission spaces of members of a social organization and restrict their activities. In Appendix, formal systems mentioned in this chapter are described in detail.

Keywords: normative system, dynamic belief-desire-obligation logic, philosophy of action, social ontology, actions in games, speech acts, role-specific normative spaces

1. Introduction

This chapter deals with some problems in social ontology. I investigate the following questions. How can we characterize agents who constitute social organizations? How is the society structured? How is the social reality created? To the first question, I answer that agents who can play team games constitute social organizations. It is my proposal to the second question that role-specific normative requirements structure social organizations. The social reality is created by agents who perform actions respecting socially accepted normative systems. This is my answer to the third question. These activities of agents can be explained by analyzing agents who play team games. We can observe that many games are clearly structured. Thus, investigations on games may contribute to analysis of the social reality.

In philosophy of action, Donald Davidson proposed to reduce intention to desire and means-end-belief [1]. Michael Bratman opposed to this claim and interpreted intention as planning [2]. Furthermore, against reductionistic approach, John Searle pointed out that some actions are strongly influenced by deontic requirements [3]. To this problem, this chapter proposes a new alternative position. Namely, I suggest taking beliefs, desires, and normative beliefs as mental bases for planning and decision-making. Some actions are temporally extended and include substructural actions. A series of actions in a game is an example of such temporally extended actions. In this chapter, I will show how beliefs, desires, and normative beliefs are connected to realize the leading desire in the given constraints.

There are three major logical approaches to investigations on dynamic inferences, namely dynamic epistemic logic [4, 5], logic of belief revision [6, 7], and discourse representation theory [8, 9]. My formal approach in this chapter is mainly related to the last two approaches. The formal details of the framework in this chapter can be found in Appendix.

For the sake of space, I do not discuss collective actions in this chapter (for collective actions, see [10]). However, the model proposed in this chapter can be also applied to characterizations of collective actions.

2. Model of agents

A game is played by agents who perform actions keeping the rule system of the game. All participating agents respect the rule system of the game when they play it. Otherwise, the game will be broken down. Then, what kind of features are required for agents who can play games? This is the leading question of this section.

A rule system of a game can be interpreted as a normative system [11]. I proposed in some articles to represent a normative system by a pair ⟨BB, OB⟩ constituted of a belief base BB and an obligation base OB [11, 12]. A player masters this normative system and decides her action based on it. Furthermore, her decision of the next action is influenced by her desire of winning the game. In each game stage, a player will perform an action type that is considered as a promising move to win the game. Here, I take a desire base DB into consideration and define Belief-Desire-Obligation system ⟨BB, DB, OB⟩ as follows [13, 14].

2.1 (S2.1) Belief-Desire-Obligation system (BDO-system)

a. I assume that each of BB, DB, OB is a set of sentences, more precisely, a set of sentences in First-Order Logic (FO-Logic). Every sentence in BB expresses what is believed, every sentence in DB expresses what is desired, and every sentence in OB expresses what is believed to be obligated. We call BB, DB, OB belief base, desire base, and obligation base respectively.

b. A pair ⟨BB, DB⟩ is a Belief-Desire system (BD-system) ⇔ BB∪DB is consistent. Here, " ⇔" is used as an abbreviation for "if and only if."

c. A pair ⟨BB, OB⟩ is a Belief-Obligation system (BO-system) ⇔ BB∪OB is consistent.

d. A triple ⟨BB, DB, OB⟩ is a Belief-Desire-Obligation system (BDO-system) ⇔ ⟨BB, DB⟩ is a BD-system and ⟨BB, OB⟩ is a BO-system.

e. A BDO system for agent A is expressed as $\langle BB_A, DB_A, OB_A\rangle$.

f. We call socially accepted BO-system normative system.

Normative systems such as laws can be expressed as BO-systems [12]. In Section 3, we will see that a rule system of a game can be interpreted as a normative system.

Note that the definition of BDO-system in (S2.1) allows conflicts between the BD-system and the BO-system. This feature reflects cases in which an agent has a desire that conflicts with her BO-system. When an agent prefers her desire even if this desire conflicts with an accepted normative system, a performance of a prohibited action becomes possible.

Now, we can introduce some logical operators and define a logical framework Belief-Desire-Obligation Logic (BDO-Logic) (for details, see Appendix (Ap.1)). For agent A, logical operators in BDO-Logic are B_A, M_A, O_A, F_A, P_A, and D_A. The intended meaning of sentences with these operators can be expressed as follows: A believes φ ($B_A\ \varphi$), A believes that possibly φ ($M_A\ \varphi$), A believes it is obligated φ ($O_A\ \varphi$), A believes it is forbidden φ ($F_A\ \varphi$), A believes it is permitted φ ($P_A\ \varphi$), and A believes it is desired φ ($D_A\ \varphi$). The following list describes some important features of these operators.

2.2 (S2.2) Operators in BDO-Logic

a. Epistemic operators B_A and M_A depend on only belief base BB_A, namely it holds: [$B_A\ \varphi \Leftrightarrow \varphi$ follows from BB_A] and [$M_A\ \varphi \Leftrightarrow \varphi$ is compatible with BB_A].

b. Normative operators O_A, F_A, and P_A depend on both belief base BB_A and obligation base OB_A. That means: [$O_A\ \varphi \Leftrightarrow$ [not $B_A\ \varphi$ and φ follows from $BB_A \cup OB_A$]], [$F_A\ \varphi \Leftrightarrow O_A \neg\varphi$], and [$P_A\ \varphi \Leftrightarrow$ [not $B_A\ \varphi$ and φ is compatible with $BB_A \cup OB_A$]].

c. Desire operators D_A is defined in analog to obligation operator O_A, namely: [$D_A\ \varphi \Leftrightarrow$ [not $B_A\ \varphi$ and φ follows from $BB_A \cup DB_A$]].

d. The inference system with B_A and M_A, which satisfies characterizations in (S2.2.a), is called B-Logic. The inference system with B_A, M_A, O_A, F_A, and P_A, which satisfies characterizations in (S2.2.a) and (S2.2.b), is called BO-Logic. The inference system with B_A, M_A, and D_A, which satisfies characterizations in (S2.2.a) and (S2.2.c), is called BD-Logic. Finally, the inference system with B_A, M_A, D_A, O_A, F_A, and P_A, which satisfies characterizations in (S2.2.a), (S2.2.b), and (S2.2.c), is called BDO-Logic.

$B_A\ \varphi$ roughly means that A believes that it must be the case φ and $M_A\ \varphi$ roughly means that A believes that it may be the case φ. Stipulation (S2.2.a) expresses a standard logical characterization of doxastic state. Characterization (S2.2.b) expresses the idea that a normative belief is formed with a presupposition of a belief base (see **Figure 1**). For example, an obligation of payment of a ticket in a museum presupposes the belief that there is such a museum. (S2.2.b) also claims that what is believed

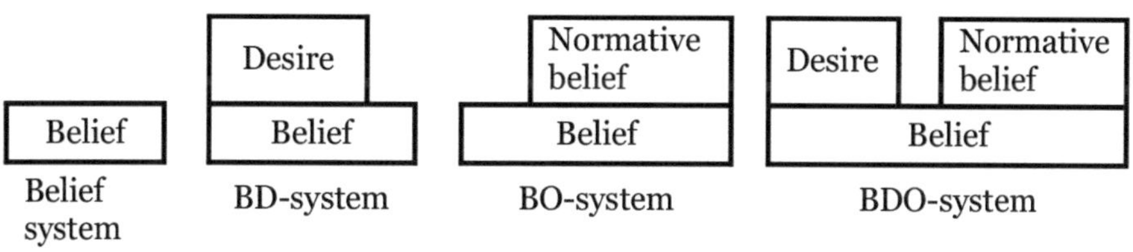

Figure 1.
Mental states of an agent (BDO-system).

cannot be normatively believed. Thus, an agent who once recognized that the required obligation was fulfilled is released from this obligation. For example, if you have already paid for a ticket, then you are released from obligation of the payment. Conforming with the standard view, (S2.2.b) characterizes permission as a deontological possibility. According to (S2.2.c), an agent who once recognized that the original desire was fulfilled will no longer try to fulfill this desire. For example, if you have already bought a book you wanted to read and recognized this result, then your desire of buying the book would disappear.

Based on BDO-Logic, I characterize intention as decision-making.

2.3 (S2.3) Characterization of four kinds of spaces for decision-making

An agent chooses an action type from a set of action sentences. In BDO-Logic, there are four kinds of such sets of action sentences (for details, see Appendix (Ap.1.i)). Here, we assume: $BDO_A = \langle BB_A, DB_A, OB_A \rangle$.

a. The desire space of agent A with BDO_A, Desire-Space(A, BDO_A), is the set of action sentences, which A desires to fulfill.

b. The obligation space of agent A with BDO_A, Obligation-Space(A, BDO_A), is the set of action sentences, which A believes being obligated to fulfill.

c. The prohibition space of agent A with BDO_A, Prohibition-Space(A, BDO_A), is the set of action sentences, which A believes being forbidden to fulfill.

d. The permission space of agent A with BDO_A, Permission-Space(A, BDO_A), is the set of action sentences, which A believes being permitted to fulfill.

e. The obligation space of agent A with BDO_A, the prohibition space of A with BDO_A, and the permission space of A with BDO_A are normative spaces of A with BDO_A.

Desire and obligation may create a direct motivation for actions, but permission has not this kind of motivational power. Rather, permission needs a support from desire. Related with normative spaces, there are three kinds of decision-making.

2.4 (S2.4) Three types of decision-making

a. [Desire-based Decision] Agent A chooses an action sentence from Desire-Space(A, BDO_A) and performs the chosen action. I call this kind of decision desire-based decision.

b. [Obligation-based Decision] Agent A chooses an action sentence from Obligation-Space(A, BDO_A) and performs the chosen action. I call this kind of decision obligation-based decision.

c. [Permission-based Decision] Agent A chooses an action sentence from Permission-Space(A, BDO_A) with help of her desire and performs the chosen action. I call this kind of decision permission-based decision.

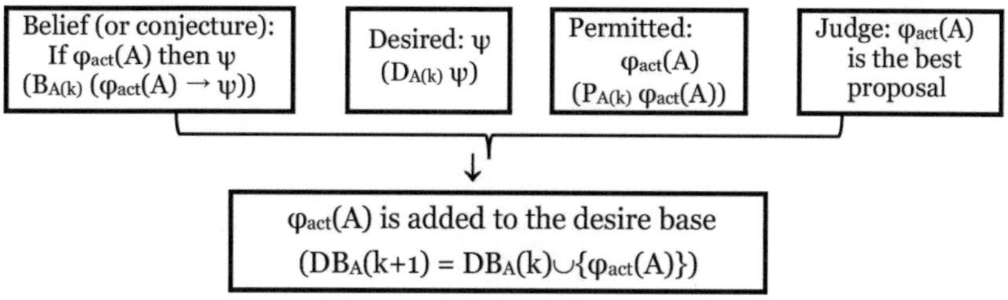

Figure 2.
Abduction-like update of desire base.

As Bratman pointed out, there might be several conflicting future options each of which is, in light of our desires and beliefs, equally attractive [15]. Characterizations in (S2.3) and (S2.4) take these phenomena into consideration. A BD-system or a BO-system provides some attractive options for decision-making. Thus. to perform an action, we need to choose one of action type from these options. Characterization (S2.4) shows that there are three types of decision-making. An agent can prefer desired-based decision or obligation-based decision or permission-based decision. In praxis, a permission-based decision is often supported by a desire-based decision (see abduction-like update of desire base in **Figure 2**).

Many philosophers in action theory focused their investigations on desire-based decisions. Searle criticized this preconception and pointed out that there are also obligation-based decisions [3]. Supplementing these preceding investigations, I will show that permission-based decisions play an important role in our everyday life.

To clarify the relationship between BDO-system and normative system, I define the notion of respect for a normative system.

2.5 (S2.5) Respect for a normative system

Here, we assume that agent A has BDO-system $\langle BB_A, DB_A, OB_A \rangle$.

a. $\langle BB_A, OB_A \rangle$ includes $\langle B_{NS}, OB_{NS} \rangle \Leftrightarrow$ (BB_{NS} is a subset of BB_A & OB_{NS} is a subset of OB_A).

b. [Compatible with a normative system] A decision that A makes is compatible with normative system $\langle BB_{NS}, OB_{NS} \rangle \Leftrightarrow$ ($\langle BB_A, OB_A \rangle$ includes $\langle B_{NS}, OB_{NS} \rangle$ & A's decision is obligation-based or permission-based).

c. [Respect] Agent A respects normative system $\langle BB_{NS}, OB_{NS} \rangle \Leftrightarrow$ any decision that A makes is compatible with $\langle BB_{NS}, OB_{NS} \rangle$.

According to (S2.5), an agent who respects a normative system obeys any obligation in it and she chooses only action types that are permitted in it. For example, a player of chess respects the normative system of chess and she plays chess keeping out of violation of it. Each time, this player tries to choose a promising move that is permitted in chess so that her desire to win will be finally fulfilled.

We can update a BDO-system $\langle BB, DB, OB \rangle$ by updating BB or DB or OB. We call the framework that allows this kind of updates Dynamic BDO-Logic (see Appendix (Ap.2)). A BDO-system in Dynamic BDO-Logic contains information about its stage.

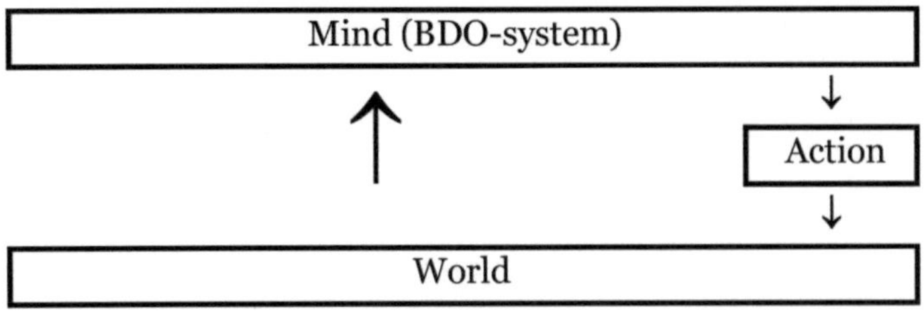

Figure 3.
Interaction between an agent and the world.

We write a BDO-system of Dynamic BDO-Logic as follows: BDO(k) = ⟨BB(k), DB(k), OB(k)⟩. A play of standard two-man games can be described in Dynamic BO-Logic that is a subsystem of Dynamic BDO-Logic [11, 13, 14]. Activities of an agent in the world can be described by using Dynamic BDO-Logic. An agent performs an action based on her current mental states. This action changes the world. When this agent recognizes this change, her mental states are updated, and she performs new action based on the new mental states. This kind of interaction between effects of action and updates of mental states goes on while an agent performs actions in the world (**Figure 3**).

According to Searle, beliefs have mind-to-world direction of fit, whereas desires and intentions have world-to-mind direction of fit [16]. This roughly means that beliefs should be formed to match the world and that the world state should be changed to match desires or intentions. This Searle's view is not false but static. When we apply the dynamic view described in **Figure 3** to analysis of a goal-directed action, we see that a belief change caused by an observation of her own first action may affect her desire and decision-making for the next action to achieve the goal.

Now, we show that rules can be analyzed by using BO-systems. According to Searle, there are two types of rules, namely regulative and constitutive rules [17]. Regulative rules regulate a pre-existing activity, an activity whose existence is logically independent of the rules. Regulative rules characteristically take the form of or can be paraphrased as imperatives, e.g., "Officers must wear ties at dinner". Constitutive rules constitute an activity the existence of which is logically dependent on the rules. Constitutive rules can be paraphrased as "X counts as Y in context C". A typical example is an introduction of a term used in a game, e.g., "A checkmate is made when the king is attacked in such a way that no move will leave it unattacked" [17]. Both rules can be expressed in BO-Logic (here, I use tr as a translation function from English into a language of FO-Logic). In the normative system for officers $\langle BB_{officer}, OB_{officer}\rangle$, tr(Officers wear ties at dinner) is a member of $OB_{officer}$. Similarly, in the normative system of chess $\langle BB_{chess}, OB_{chess}\rangle$, tr(A checkmate is made if and only if the king is attacked in such a way that no move will leave it unattacked) is a member of BB_{chess}.

3. Description of games

What is a game? I require the following two conditions for games.

(S3.1.a) There is a normative system for the game that all participants of the game accept and respect.

(S3.1.b) The start and the end condition of the game are clearly characterized in the normative system of the game.

These requirements are not demanding. The first requirement (S3.1.a) is essential for my enterprise. Following this requirement, I interpret a rule system of a game as a normative system. The start and the end condition in the second requirement (S3.1.b) are required to temporally demarcate a game.

A play of standard two-man games can be described in Dynamic BDO-Logic. In Subsection 3.1 and Appendix (Ap.3)-(Ap.5), by using example of Tic-Tac-Toe, I demonstrate how to express its rule system as a normative system. More complex games and team games can be also expressed as normative systems [11]. In Subsection 3.3, I demonstrate how to represent team games by using Dynamic BDO-Logic.

3.1 Description of simple games

Tic-Tac-Toe is a paper-and-pencil game for two players who take turns marking the spaces in a 3 × 3 grid. The player who succeeds in placing three of her marks in a horizontal, vertical, or diagonal row wins the game (the discussion in this subsection is largely based on [13]). There are two types of propositions in Tic-Tac-Toe, namely action types and state types (for state types, see Appendix (Ap.3)). There is only one action type in Tic-Tac-Toe, namely: X places her mark in position s in stage k. The normative system for Tic-Tac-Toe is $\langle BB_{ttt}, OB_{ttt} \rangle$. To play Tic-Tac-Toe, every player must master this normative system. Contents of BB_{ttt} and OB_{ttt} can be explained as follows (for details, see Appendix (Ap.3) and (Ap.4)).

(S3.2) Belief base BB_{ttt} consists of the following elementary theory for Tic-Tac-Toe.

a. There are exactly two players who are opponents each other.

b. We use O and × as two marks of players.

c. If one player wins in Tic-Tac-Toe, then her opponent loses.

d. When it is turn of a player, it is not turn of her opponent.

e. If position s is occupied by mark m in stage k and $k \leq n$, then s is still occupied by m in stage n.

f. Position s is vacant in stage k ⇔ s in not occupied by any mark in stage k.

g. The player who succeeds in placing three of her marks in a horizontal, vertical, or diagonal row wins the game.

h. The game ends ⇔ a player wins or there is no vacant position in the 3 × 3 grid.

i. If a player places his mark m in position s in stage k, then s is occupied by m in k + 1 and the turn is alternated in k + 1.

(S3.3) Obligation base OB_{ttt} consists of the following sentences.

a. If the game is not yet ended and it is X's turn in stage k, then X places her own mark into a vacant position in k.

b. Any player does not place her mark into a non-vacant position in any stage.

c. Any player does not place her mark when it is not her turn.

d. Any player does not write her mark when the game ends.

In Tic-Tac-Toe, there are two players X and Y. In many two-person games, their rule systems and the game developments are shared by players. Let X + Y be the group consisting of X and Y. Then, a shared BDO-system can be defined as follows: $\langle BB_{X+Y}(k), DB_{X+Y}(k), OB_{ttt}\rangle$ with $BB_{X+Y}(k) = BB_X(k) \cap BB_Y(k)$ and $DB_{X+Y}(k) = DB_X(k) \cap DB_Y(k)$. Thus, $BB_{X+Y}(k)$ is the shared belief base of X + Y in stage k, $DB_{X+Y}(k)$ is the shared desire base of X + Y in stage k, and OB_{ttt} is the shared obligation base of X + Y. Initial-state is the set of sentences that describe the initial state and contains sentences such as "Every position (in 3×3 grid) is vacant in stage 0". We assume that $BB_{X+Y}(0)$ is BB_{ttt} supplemented by Initial-state (see Appendix (Ap.4.b) and (Ap.4.c)). As an abbreviation, we write: $BDO_{X+Y}(k) = \langle BB_{X+Y}(k), DB_{X+Y}(k), OB_{ttt}\rangle$. During the play of the game, $BB_{X+Y}(k)$ is updated by adding acquired new information to the previous belief base $BB_{X+Y}(k - 1)$ (see (S3.4.a) and Appendix (Ap.4.f1)).

Note that (S3.3.a) and (S3.3.c) play an important role for the calculation of obligation space. By the way, it holds in BDO-Logic: $(O(\varphi \rightarrow \psi) \,\&\, B(\varphi)) \Rightarrow O(\psi)$ (see [11] and Appendix (Ap.1.h)). Then: $O_{X+Y}(k)\ (\varphi \rightarrow \psi) \Rightarrow (B_{X+Y}(k)\ \varphi \Rightarrow O_{X+Y}(k)\ \psi)$. Now, it holds because of (S3.3.a): Players X + Y believe in stage k that it is obligated that [if the game is not yet ended and it is X's turn in k, then X places her own mark into a vacant position in k]. Thus, it holds: When players X + Y know that the game is not yet ended and it is X's turn in stage k, they believe that it is obligated that X places her own mark into a vacant position in k (this sentence has the following form: $B_{X+Y}(k)\ \varphi \Rightarrow O_{X+Y}(k)\ \psi$). Thus, when X + Y know that the game is not yet ended and it is X's turn in k, Obligation-Space(X, $BDO_{X+Y}(k)$) is the set consisting of tr(X places her mark into a vacant position in stage k). Similarly, because of (S3.3.c), when X + Y know that it is not Y's turn in stage k, then Obligation-Space(Y, $BDO_{X+Y}(k)$) is the set consisting of tr(Y does not place her mark in any position in stage k).

To play Tic-Tac-Toe seriously, it is required that players desire to play it and to win (thus, for every player X, $DB_X = \{play(X, Tic\text{-}Tac\text{-}Toe), \exists k\ won(X, k)\}$). Because a player desires to win a game, she tries to find good strategies for the victory.

There are two types of stages, namely, continuing stage and terminal stage. The game process of Tic-Tac-Toe depends on these two qualifications of stages. Here, we describe two schemas for updates of a BDO-system.

(S3.4.a) [Continuing stage] In stage k, the following beliefs are shared by all players: The game is not ended in k, it is X's turn in k, and Y is X's opponent. Then, it follows that Obligation-Space(X, $BDO_{X+Y}(k)$) = {tr(X places her mark in a vacant position in k)}, Obligation-Space(Y, $BDO_{X+Y}(k)$) = {tr(Y does not place her mark in any position in k)}, Permission-Space(X, $BDO_{X+Y}(k)$) is the set of sentences that express permissible placing actions in k, and Permission-Space(Y, $BDO_{X+Y}(k)$) is empty. Here, X wants to win the game and X chooses a placing action type from Permission-Space(X, $BDO_{X+Y}(k)$) to fulfill her obligation and to approach her leading desire. Now, suppose that X forms the desire to perform a placing action expressed by $\varphi_1(X)$. Thus, $DB_{X(k)}$ is obtained by adding $\varphi_1(X)$ to $DB_{X(k-2)}$. After X's performance of this action, both players confirm it and update their belief base so that it holds: $BB_{X+Y}(k+1)$ is the union of $BB_{X+Y}(k)$ and $\{\varphi_1(X)\}$. During the play, the number of vacant positions in the grid diminishes, and the permission spaces of both players decreases. (In detail, see Appendix (Ap.4.f1))

(S3.4.b) [Terminal stage] In stage k, all players believe that the game ended in k. In this case: Permission-Space of X and Y in k are empty, so that they stop to play. (In detail, see Appendix (Ap.4.f2))

Here, let us give an example of a play of Tic-Tac-Toe. In this example, each of A and B wants to play Tic-Tac-Toe and wants to win.

Suppose that A starts with placing his mark O in position [1,1] of the grid (formally: placing(A, O, [1,1], 0)). Both players try to win and make a promising move in each game stage. The following sequence of quadruples describes one possible game development: ⟨(A, O, [1,1], 0), (B, ×, [2,2], 1), (A, O, [3,3], 2), (B, ×, [1,3], 3), (A, O, [3,1], 4), (B, ×, [3,2], 5), (A, O, [2,1], 6)⟩. After A placed O into position [2,1] in stage 6, A and B recognize in stage 7 that three of O are in a vertical raw and that A won the game. Results of this action sequence is partly demonstrated in **Figure 4**. Here, we can see that the terminal stage of this game has been achieved in stage 7.

Now, let us closely investigate the relation between desire and permission in this game process. In the initial stage, A desires to win the game and A believes that A is permitted to place O in any position in 3 × 3 grid. Winning this game is the final desire of A in this game. In each stage, when it is A's turn, A forms a desire to perform a placing action that is permitted and will lead to fulfillment of her final desire. Update of desire during a game can be interpreted as an abduction-like reasoning demonstrated in **Figure 2**, where $\varphi_{act}(A)$ expresses A's performance of a placing action (an abduction (inference to the best explanation) can be demonstrated as **Figure 5**). In stage 0, A desires to place O into [1,1], because A conjectures that it can lead to A's victory. In stage 6, A believes that placing O into [2,1] is the best action to A's victory. Based on this consideration, A desires to perform this action and performs it. A's intention to play Tic-Tac-Toe corresponds to the prior intention and A's intention to perform the chosen placing action corresponds to intention-in-action in the literature [16].

3.2 Speech acts and creation of social facts

Many team games are controlled by referees, and statements made by them play an important role in game processes. In team games, referees have authority, and their

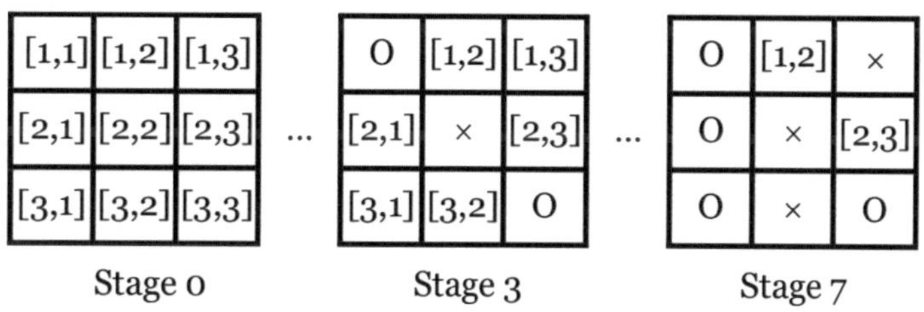

Figure 4.
An example of game progresses in Tic-Tac-Toe.

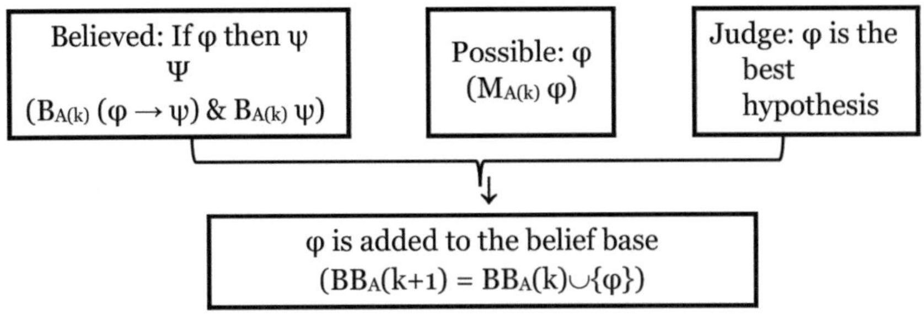

Figure 5.
Schema for abduction.

judgments are accepted by players and audiences. A judgment by a referee can be characterized as a declaration that is a type of illocutionary acts. In this subsection, to characterize declarations, we investigate speech acts based on Dynamic BDO-Logic. For this purpose, let us start with Searle's characterizations of five types of illocutionary acts. They are Assertives, Directives, Commissives, Expressives, and Declarations, which Searle characterizes as follows [18].

(S3.5.a) [Assertives] The point or purpose of the members of the assertive class is to commit the speaker (in varying degrees) to something's being the case, to the truth of the expressed proposition. Examples for assertives are assertion, complaint, conclusion, deduction, and so on.

(S3.5.b) [Directives] The illocutionary point of these consists in the fact that they are attempts (of varying degrees, and hence, more precisely, they are determiners of the determinable, which includes attempting) by the speaker to get the hearer to do something. Examples for directives are command, request, permission, advise, invitation, and so on.

(S3.5.c) [Commissives] Commissives are those illocutionary acts whose point is to commit the speaker (again in varying degrees) to some future course of action. Examples for commissives are promise, pledge, and so on.

(S3.5.d) [Expressives] The illocutionary point of expressives is to express the psychological state specified in the sincerity condition about a state of affairs specified in the propositional content. Examples for expressives are gratitude, apology, congratulation, and so on.

(S3.5.e) [Declarations] It is the defining characteristic of declarations that the successful performance of one of its members brings about the correspondence between the propositional content and reality, successful performance guarantee that the propositional content corresponds to the world. Examples for declarations are appointment, naming, and so on.

I propose to interpret these illocutionary acts as actions that are aimed to update belief base or obligation base.

(S3.6) Updates of BDO-bases by performances of illocutionary acts.

We assume: S is a speaker, H is a hearer, S + H is the group consisting of S and H, and G(S) is a group including S as a member. Let $\varphi_{act}(X)$ be an action sentence with X as the agent of this action. Furthermore, let BB_{S+H}(now) be the current belief base of S + H, OB_{S+H}(now) be the current obligation base of S + H, and $BB_{G(S)}$(now) be the current belief base of group G(S). Here, I propose that a successful performance of an illocutionary act effects the following updates of BDO-bases.

a. [Assertives] If S successfully performs an assertive illocutionary act with content expressed by φ addressing to H, then φ is added to BB_{S+H}(now).

b. [Directives] If S successfully performs a directive illocutionary act with content expressed by φ_{act}(H) addressing to H, then φ_{act}(H) is added to OB_{S+H}(now).

c. [Commissives] If S successfully performs a commissive illocutionary act with content expressed by φ_{act}(S) addressing to H, then φ_{act}(S) is added to OB_{S+H}(now).

d. [Expressives] If S successfully performs an expressive illocutionary act with content expressed by φ addressing to H, then φ is added to BB_{S+H}(now).

e. [Declarations] If S successfully performs a declaration with content expressed by φ addressing to G(S), then φ is added to $BB_{G(S)}$(now).

f. [Basic social facts] A sentence φ expresses a basic social fact for group G ⇔ φ is true because G has accepted φ (Thus, φ would not be true, if G had not accepted φ).

g. [Social facts] A sentence φ expresses a social fact for group G ⇔ φ is true because there is a basic social fact for G that supports the truth of φ. Thus, a basic social fact is also a social fact.

Speech act theory of John Austin and Searle investigates only singular sentences. This marks a limitation of this theory. Normally, utterances are made as a part of a comprehensive action. For example, in a discussion in a group, many assertions and many questions are made to reach an agreement. This kind of discussions can be considered as a game. The game starts with an identification of the goal of the discussion, and it ends when this goal is achieved. A contribution to the discussion can be considered as a move in a game. This game successfully ends when participants achieved an acceptable conclusion, otherwise it unsuccessfully ends. According to (S3.6.f) and (S3.6.g), when the discussion group G reaches an agreement expressed by φ, then the fact expressed by φ becomes a social fact in G. From the same reason, a successful declaration introduces a new social fact. Later, the so created social facts may become bases for individual and collective actions of members of G.

3.3 Description of team games

Team games can be also described in Dynamic BDO-Logic or in its subsystem [11]. For this description, expressions of roles are crucial. For there are many obligations that are imposed only to agents who have certain roles. This role-specific obligation can be expressed as an element of obligation base as follows: If x has $role_1$, then x performs $action_1$. For example, a goalkeeper in soccer has the following obligation: If agent x is a goalkeeper, then x tries to stop the ball from going into x's team's goal. There are also goalkeeper-specific permissions. For example, it is permitted only for a goalkeeper to hold the ball with her hands within team's penalty box. In other words, it is forbidden for all players other than goalkeepers to use her hands and it is forbidden for a goalkeeper to hold the ball with her hands outside of team's penalty box. As these examples show, roles in games can be formally captured by determining role-specific obligation space and permission space. The rule system for soccer can be expressed as a normative system $\langle BB_{soccer}, OB_{soccer}\rangle$, which determines role-specific normative spaces.

Let me explain these role-specific normative spaces by using an example of baseball games. In baseball, there are two main normative systems, namely a normative system for players and a normative system for umpires, represented as $\langle BB_{bb\text{-}players}, OB_{bb\text{-}players}\rangle$ and $\langle BB_{bb\text{-}umpires}, OB_{bb\text{-}umpires}\rangle$ respectively. Pitcher, catcher, first baseman, and so on belong to player-roles for defense, whereas there is only one player-role for offense, namely batter.

All players in a team desire their team's victory. Thus, it holds that all members of team G_1 have desire to win against team G_2. Thus, for every member X of G_1, X's desire base contains tr(G_1 win the game). This desire for a victory of own team is for its members the leading desire during the game. To fulfill this goal, players perform their actions respecting their normative spaces. When a player keeps permission space, she automatically keeps out of prohibition space. In general, given normative system $\langle BB_{NS}, OB_{NS}\rangle$, three kinds of normative spaces can be characterized as follows (for details, see Appendix (Ap.6)).

> (S3.7.a) Obligation-Space*(A, $role_1$, G, t, $\langle BB_{NS}, OB_{NS}\rangle$) is the set of action sentences, which $\langle BB_{NS}, OB_{NS}\rangle$ obligates agent A to fulfill when A has $role_1$ in group G at time t.
>
> (S3.7.b) Prohibition-Space*(A, $role_1$, G, t, $\langle BB_{NS}, OB_{NS}\rangle$) is the set of action sentences from the realization of which $\langle BB_{NS}, OB_{NS}\rangle$ prohibits agent A when A has $role_1$ in group G at time t.
>
> (S3.7.c) Permission-Space*(A, $role_1$, G, t, $\langle BB_{NS}, OB_{NS}\rangle$) is the set of action sentences, which $\langle BB_{NS}, OB_{NS}\rangle$ allows agent A to fulfill only when A has $role_1$ in group G at time t.
>
> (S3.7.d) (S3.7.a), (S3.7.b), and (S3.7.c) characterize role-specific obligation, prohibition, and permission spaces. In this chapter, we call these spaces roll-specific normative spaces.

Rules of baseball can be divided into declarative and normative sentences. The FO-translations of declarative sentences are assembled into $BB_{bb\text{-}players}$. The normative sentences are reformed in obligation sentences, and their FO-translations are assembled into $OB_{bb\text{-}players}$ (see [11]). Then, based on (S3.7), we can calculate role-specific obligation and permission spaces for all players. For example, the sentence tr(A stations herself directly back of the plate) belongs to Obligation-Space*(A, catcher, G_1, t, $\langle BB_{bb\text{-}players}, OB_{bb\text{-}players}\rangle$), when A is a catcher in team G_1 and t belongs to the defense time for G_1. A baseball game will be played without any problems, when all players respect $\langle BB_{bb\text{-}players}, OB_{bb\text{-}players}\rangle$ and try to fulfill their role-specific normative requirements.

There is also a normative system for umpires. For example, if the chief umpire of a baseball game believes in time t that the ball that the pitcher has thrown is a strike, then tr(A declares the thrown ball as a strike) belongs to Obligation-Space*(A, chief umpire, $G_{umpires}$, t, $\langle BB_{bb\text{-}umpires}, OB_{bb\text{-}umpires}\rangle$).

Let $G_{audience}$ denote the group of audiences of this game and assume that G_{all} consists of $G_{audiences}$, $G_{players}$, and $G_{umpires}$. In a baseball game, G_{all} accepts every declaration that umpires make in this game. For example, if an umpire declares φ, then φ is added to the belief base of G_{all} so that φ becomes a sentence that expresses a social fact for G_{all}. The fact expressed by φ is never doubted during the game and it is shared by all members of G_{all}. The BO-system for G_{all} at t can be expressed as $\langle BB_{G(all)}(t), OB_{G(all)}(t)\rangle$, where $BB_{G(all)}(t)$ contains both of $BB_{bb\text{-}players}$ and descriptions of all events in the game unitl t. For players and the manager of a team, knowledge of social facts during the game provides a basis for planning of next actions in the game. In other words, they plan next actions respecting role-specific normative spaces at the given situation.

In the philosophical literature, many philosophers have required mutual beliefs as a basis for collective actions [19, 20]. However, this requirement is too strong, as some philosophers pointed out [3, 21]. Therefore, I propose the following tic-for-tac-like strategy as an alternative, and I call this strategy "acceptance strategy".

(S3.8.a) At the start of a game, you assume that all players accept and respect the normative system of the game.

(S3.8.b) You respect the normative system of the game so long as you do not find any violation of the normative system by any player.

This strategy is strong enough for assuring the play, and it is far weaker and more natural than the requirement of mutual beliefs among players.

4. Games and society

Max Weber, a famous sociologist, stated that the interpretative understanding of social actions is an aim of sociology [22]. In this section, I propose that social actions should be understood in a wide sense. For not only collective actions but also individual actions can be considered as social actions when they presuppose socially accepted normative systems. For example, a violation of a criminal law is only possible when a socially accepted criminal law exists in that society. In this sense, a criminal action by an individual agent can be considered as a social action insofar she is aware that her action violates a criminal law. In other words, each citizen has a prohibition space that forbids criminal actions, while a criminal person intentionally ignores this space.

In the society, there are many social organizations, such as states, armed forces, companies, universities, hospitals, and so on. Each of these social organizations has its own normative system and most of its members respect it. Some social organizations such as states have a set of complex normative systems, and its members only partially know about them. At any rate, mature members of a social organization share some parts of its normative system.

Here, I distinguish two types of shared BO-systems, namely individually shared and socially shared BO-system.

(S4.1.a) [Shared BO-system] We assume that A has BO-system $\langle BB_A, OB_A\rangle$ and that B has BO-system $\langle BB_B, OB_B\rangle$. Then, A and B share BO-system $\langle BB, OB\rangle \Leftrightarrow$ [$\langle BB_A, OB_A\rangle$ includes $\langle BB, OB\rangle$ & $\langle BB_B, OB_B\rangle$ includes $\langle BB, OB\rangle$].

(S4.1.b) [Socially shared BO-system] A and B socially share ⟨BB, OB⟩ ⇔ There is a group G such that [G is greater than A + B & members of G accept ⟨BB, OB⟩ as a normative system & A and B share ⟨BB, OB⟩].

(S4.1.c) [Individually shared BO-system] A and B individually share ⟨BB, OB⟩ ⇔ [A and B share ⟨BB, OB⟩ & A and B do not socially share ⟨BB, OB⟩].

In many social organizations, certain roles are assigned to their members, and they perform actions within their role-specific normative spaces to achieve a shared goal or an individual goal. In such cases, socially shared normative systems construct bases for role-specific normative spaces (see (S3.7.a)-(S3.7.d) and Appendix (Ap.6)). These socially shared normative systems are not originally founded by isolated individuals. Contrarily, individuals acquire them as already socially accepted systems. The reality of a social organization is founded on this socially shared normative systems and role-specific normative spaces that motivate or restrict actions of their members.

There are many interconnected normative systems in a society [11]. There are also fundamental normative systems, such as a monetary system, that support many social activities. For example, economic activities are only possible in a society where a monetary system is accepted and functioning. This monetary system is a normative system that can be described as $\langle BB_{ms}, OB_{ms}\rangle$. Many mature agents in this society have BO-systems, which include $\langle BB_{ms}, OB_{ms}\rangle$. Most members of the society accept this monetary system, restrict their desires by it, and make their decision within the normative spaces determined by it. The civil law and the corporation law contain many texts that are related to a monetary system. These laws in a country should be consistent as normative systems, where the consistency of normative systems is defined as follows.

(S4.2) Normative systems $\langle BB_1, OB_1\rangle, \ldots, \langle BB_k, OB_k\rangle$ are consistent ⇔ $\langle BB_{all}, OB_{all}\rangle$ is consistent in sense of (Ap.1.a) in Appendix, where BB_{all} is the union of $BB_1, \ldots, BB_k$ and OB_{all} is the union of $OB_1, \ldots, OB_k$.

An agent might be simultaneously associated with different roles in different organizations. In such a case, she has many different role-specific normative spaces. In other words, there are at least two groups G_1, G_2, and normative systems $\langle BB_{NS1}, OB_{NS1}\rangle$, $\langle BB_{NS2}, OB_{NS2}\rangle$ such that A has, in time t, two different roll-specific obligation spaces, namely Obligation-Space*(A, $role_1$, G_1, t, $\langle BB_{NS1}, OB_{NS1}\rangle$) and Obligation-Space*(A, $role_2$, G_2, t, $\langle BB_{NS2}, OB_{NS2}\rangle$). For example, the President of Liberal Democratic Party (LDP) in Japan is often the same person with the Prime Minister in Japan. Such a person A simultaneously has two different role-specific normative spaces, namely Obligation-Space*(A, president, LDP in Japan, t, $\langle BB_{NS(LDP)}, OB_{NS(LDP)}\rangle$) and Obligation-Space*(A, prime minister, Japan, t, $\langle BB_{Japanese(Laws)}, OB_{Japanese(Laws)}\rangle$).

Many social actions can be interpreted as moves in a game because they presuppose certain normative systems. When these actions constitute parts of a comprehensive action that has a start stage and an end stage, they can be interpreted as moves in a game in sense of (S3.1.a) and (S3.1.b). A construction of a building, an execution of a project, and so on are such comprehensive actions that can be interpreted as team games that involve division of labor.

Social fact and division of labor are two kye concepts of sociologist Émil Durkheim [23]. I have already given a characterization of social facts in (S3.6.g). Here, I characterize notion of division of labor. The division of labor is used for construction of a social organization. For example, a university is divided into many departments through division of labor. Role-specific normative spaces play an important role in the division of labor. As an example, let us describe the structure of the Cabinet of Japan. The Cabinet of Japan consists of a Prime Minister and up to nineteen Ministers of States due to the Constitution of Japan. Under the Constitution, the Prime Minister has distinguished powers. For example, the Prime Minister can decide the dissolution of the House of Representatives. This power can be expressed as follows: tr(A decides the dissolution of the House of Representatives in t) belongs to Permission-Space*(A, prime minister, Japan, t, $\langle BB_{constitution(Japan)}, OB_{constitution(Japan)}\rangle$), and for any person B who is not the Prime Minister, tr(B decides the dissolution of the House of Representatives in t) belongs to Prohibition-Space*(B, not(prime minister), humans, t, $\langle BB_{constitution(Japan)}, OB_{constitution(Japan)}\rangle$). Naturally, there are also many obligations that are only imposed on the Prime Minister. The sentences that express these obligations belong to Obligation-Space*(A, prime minister, Japan, t, $\langle BB_{constitution(Japan)}, OB_{constitution(Japan)}\rangle$).

What does it mean that a social norm is accepted in a society? Some philosophers, like David Lewis, Bratman, and Raimo Tuomela, require mutual beliefs among members for collective acceptance and agreement [19, 20, 24]. However, it is often criticized that this requirement is too strong [3, 21], and I agree with this criticism. Thus, in analog to acceptance strategy for normative systems of games (see (S3.8.a) and (S3.8.b) in Subsection 3.3), I propose the following slightly weekend acceptance strategy for social normative systems.

(S4.3.a) At the start, you assume that most of members in the organization respect the social normative system.

(S4.3.b) You respect the normative system so long as you do not find many violations of it by other members.

A social normative system loses its effectivity in a social organization when many members of the organization perform actions that it prohibits. They violate the social normative system and choose actions that lie outside of its permission space. This sometimes happens in a social crisis such as in a financial panic or in a political revolution. The acceptance strategy explains why this kind of social collapses is possible. In a social crisis, many members of the society prefer their own life maintenance to sustainability of the society. If many members of the society follow a normative system based on the acceptance strategy (S4.3.b), then they will cease to respect it when they see that many members violate it.

5. Conclusions

In this chapter, I introduced a formal framework, Dynamic Belief-Desire-Obligation Logic (Dynamic BDO-Logic), and I demonstrated how to describe two-player games and team games within this framework. Team games provide concrete examples for study of agent's individual and collective actions under acceptance of a normative system.

Many social activities of agents can be interpreted as moves in a game. The social reality is created by agents who plan their actions based on social normative systems and perform them. The structure of the society is constituted by acceptance of roles.

An agent in a social organization accepts her role in it and performs actions within role-specific normative spaces. These kinds of activities of agents support the persistence of a society.

A human is born in a society that is already structured by activities of its current members and forerunners. She can choose her future actions, but her choice is always restricted by existing social normative systems. It will be an illusion if one tries to construct a society solely from an agreement among totally free isolated agents.

Acknowledgements

I would like to thank the editor, Dr. Branislav Sobota, for valuable suggestions on the earlier draft of this chapter.

Appendix

In this appendix, I precisely describe the definitions and the theorems mentioned in the main text.

(Ap.1) Definition and Characterization of BDO-Logic

We use not, &, or,⇒, and ⇔ as meta-language expressions of logical connectives. Let Cn(X) be an abbreviation for the deductive closure of X, where X is a set of FO-sentences. Thus, Cn(X) contains every FO-sentence that can be logically deduced from Cn(X). Furthermore, for set X of FO-sentences, we define the consistency of X as follows: consistent(X) ⇔ there is no FO-sentence φ such that $\varphi \in Cn(X)$ and $\neg\varphi \in Cn(X)$. Here, we assume that each of BB, DB, and OB is a set of FO-sentences.

a. [BO-system] Pair ⟨BB, OB⟩ is a BO-system ⇔ consistent(BB∪OB).

b. [BD-system] Pair ⟨BB, DB⟩ is a BD-system ⇔ consistent(BB∪DB).

c. [BDO-system] Triple ⟨BB, DB, OB⟩ is a BDO-system ⇔ [⟨BB, DB⟩ is a BD-system & ⟨BB, OB⟩ is a BO-system].

d. In the following definitions, we assume that agent A has BDO-system BDO_A, where $BDO_A = \langle BB_A, DB_A, OB_A \rangle$. A sentence in BDO-Logic is recursively defined as follows.

(d1) If φ is a FO-sentence and BDO_A is a BDO-system, then each of $B_A\,\varphi$, $M_A\,\varphi$, $O_A\,\varphi$, $F_A\,\varphi$, $P_A\,\varphi$, and $D_A\,\varphi$ is a sentence in BDO-Logic.

(d2) If each of Φ and Ψ is a sentence in BDO-Logic, then each of not Φ, Φ & Ψ, Φ or Ψ, $\Phi \Rightarrow \Psi$, and $\Phi \Leftrightarrow \Psi$ is a sentence in BDO-Logic.

(d3) Every sentence in BDO-Logic satisfies (Ap.1.d1) or (Ap.1.d2).

e. Now, we define operators B_A, M_A, O_A, F_A, P_A, and D_A as abbreviations in meta-language.

(e1) [Belief] $B_A\varphi \Leftrightarrow \varphi \in Cn(BB_A)$.

(e2) [Possibility] $M_A\varphi \Leftrightarrow$ consistent$(BB_A \cup \{\varphi\})$.

(e3) [Obligation] $O_A\varphi \Leftrightarrow [\varphi \in Cn(BB_A \cup OB_A)$ & not $(\varphi \in Cn(BB_A))]$.

(e4) [Prohibition] $F_A\varphi \Leftrightarrow O_A\neg\varphi$.
(e5) [Permission] $P_A\varphi \Leftrightarrow$ [consistent($BB_A \cup OB_A \cup \{\varphi\}$)& not ($\varphi \in Cn(BB_A)$)].
(e6) [Desire] $D_A\varphi \Leftrightarrow$ [$\varphi \in Cn(BB_A \cup DB_A)$ & not ($\varphi \in Cn(BB_A)$)].

f. [Truth] Sentence Φ in BDO-Logic is true in $BDO_A \Leftrightarrow \Phi$ follows from (Ap.1.e1)-(Ap.1.e6).

g. [Validity] Sentence Φ in BDO-Logic is valid $\Leftrightarrow \Phi$ is true in all BDO-systems.

h. [Theorem] There are many sentences in BDO-Logic that are valid [11]. For example, $(O_A(\varphi \rightarrow \psi)$ & $B_A\,\varphi) \Rightarrow O_A\,\psi$ is a valid sentence in BDO-Logic. The proof is straightforward. We assume $O_A(\varphi \rightarrow \psi)$ & $B_A\,\varphi$. Then, because of (Ap.1.e1) and (Ap.1.e3), $\psi \in Cn(BB_A \cup OB_A)$. If $\psi \in Cn(BB_A)$, then $(\varphi \rightarrow \psi) \in Cn(BB_A)$, which contradicts to $O_A(\varphi \rightarrow \psi)$. Thus, not ($\psi \in Cn(BB_A)$). Then, because of (Ap.1.e3), $O_A\,\psi$. Q.E.D.

i. An agent chooses an action type from a set of action sentences. In BDO-Logic, there are four kinds of such sets of action sentences. Here, we assume that $\varphi_{act}(A)$ is a FO-sentence that expresses that A performs certain action.

(i1) Desire-Space(A, BDO_A) = $\{\varphi_{act}(A): D_A\,\varphi_{act}(A)\}$.
(i2) Obligation-Space(A, BDO_A) = $\{\varphi_{act}(A): O_A\,\varphi_{act}(A)\}$.
(i3) Prohibition-Space(A, BDO_A) = $\{\varphi_{act}(A): F_A\,\varphi_{act}(A)\}$.
(i4) Permission-Space(A, BDO_A) = $\{\varphi_{act}(A): P_A\,\varphi_{act}(A)\}$.

(Ap.2) Dynamic Belief-Desire-Obligation Logic (Dynamic BDO-Logic)

We can update a BDO-system ⟨BB, DB, OB⟩ by updating BB or DB or OB. We call the framework that allows this kind of updates Dynamic BDO-Logic. A BDO-system in Dynamic BDO-Logic contains information about its stage. We write a BDO-system of Dynamic BDO-Logic as follows: BDO(k) = ⟨BB(k), DB(k), OB(k)⟩.

(Ap.3) Formalization of Games (Example: Tic-Tac-Toe)

Here, by using example of Tic-Tac-Toe, I demonstrate how Dynamic BDO-Logic can be applied to descriptions of game processes. The normative system for Tic-Tac-Toe is a BO-system $\langle BB_{ttt}, OB_{ttt}\rangle$. There are two types of propositions, namely action tapes and state types:

Action type: placing(X, mark(X), s, k).
State type: occupied(s, m, k), turn(X, s), vacant(s, k), opponent(X, Y), end(k), won(X, k), play(X, g).

Here, X and Y are used as variables of players, s, s_1, ... are used as variables of positions, m is used as a variable of marks, k and n are used as variables of game stages, and g is a variable for games. For the sake of readability, I use Many-Sorted Logic instead of FO-Logic. However, all formulas of Many-sorted Logic can be translated into formulas of FO-Logic. I use a function mark(X), where there are two values for this function, namely O and $\times$.

a. Belief base BB_{ttt} consists of the following elementary theory for Tic-Tac-Toe: BB_{ttt} = {(ET1), (ET2), (ET3), (ET4), (ET5), (ET6)}.

(ET1) There are some trivial stipulations that can be expressed in FO-Logic. For the sake of understandability, we verbally summarize these conditions as follows: (1) There are exactly two players who are opponents each other. (2) We use O and × as two marks of players (∀X (mark(X) = O ∨ mark(X) = ×). (3) If one player wins in Tic-Tac-Toe, then her opponent loses. (4) When it is turn of a player, it is not turn of her opponent.

(ET2) [Persistence] ∀k ∀n ∀s ∀m (occupied(s, m, k) ∧ k ≤ n → occupied(s, m, n)).
(ET3) ∀k ∀s (vacant(s, k)↔ ¬ ∃ m occupied(s, m, k)).
(ET4) The definition of victory in Tic-Tac-Toe can be given in FO-Logic. However, for the sake of understanderbility, we verbally express the condition for victory: The player who succeeds in placing three of their marks in a horizontal, vertical, or diagonal row wins the game.
(ET5) ∀k (end(k) ↔ (∃X won(X, k) ∨¬ ∃ s vacant(s, k)).
(ET6) [Effect] ∀k ∀X ∀Y ∀s ((placing(X, mark(X), s, k) ∧ opponent(X, Y)) → (occupied(s, mark(X), k + 1) ∧ turn(Y, k + 1))).

b. Obligation base OB_{ttt} is defined as follows: OB_{ttt} = {(OB1), (OB2), (OB3), (OB4)}.

(OB1) ∀X ∀k (¬ end(k) ∧ turn(X, k) → $\exists^{=1}$s (placing(X, mark(X), s, k) ∧ vacant (s, k))).
(OB2) ∀X ∀k ¬∃ s (placing(X, mark(X), s, k) ∧¬ vacant(s, k)).
(OB3) ∀X ∀k (¬ turn(X k) → ¬ ∃ s placing(X, mark(X), s, k)).
(OB4) ∀X ∀k (end(k) → ¬ ∃ s placing(X, mark(X), s, k)).

c. The normative system for Tic-Tac-Toe is $\langle BB_{ttt}, OB_{ttt}\rangle$.

(Ap.4) By using Dynamic BDO-Logic, game processes of Tic-Tac-Toe can be described. Here, we describe game processes of Tic-Tac-Toe in general.

a. [Abbreviation 1] PLACING(X, mark(X), {s_1, ..., s_n}, k) = {placing(X, mark(X), s_1, k), ..., placing(X, mark(X), s_n, k)}.

b. Initial-state = {∀s vacant(s, 0), List(o) = {[1,1], [1,2], [1,3], [2,1], [2,2], [2,3], [3,1], [3,2], [3,3]}, mark(X) = O, mark(Y) = ×, turn(X, 0)}.

c. $BB_X(o) = BB_{ttt} \cup$Initial-state & DB_X = {play(X, Tic-Tac-Toe), ∃k won(X, k)} for X∈{A, B}.

d. [BDO-systems] $BDO_X(k) = \langle BB_X(k), DB_X(k), OB_{ttt}\rangle$ &
$BDO_{X+Y}(k) = \langle BB_X(k) \cap BB_Y(k), DB_X(k) \cap DB_Y(k), OB_X \cap OB_Y\rangle$. Thus, it holds: $OB_X \cap OB_Y = OB_{ttt}$.

e. Obligation space and permission space.

(e1) Obligation-Space(X, $BDO_{X+Y}(k)$) = {$\exists^{=1}$s (placing(X, mark(X), s, k) ∧ vacant(s, k)): $B_{X(k)}$ (¬end(k) ∧ turn(X, k))}. This can be proved based on (Ap.1.i2) and (OB1).
(e2) Obligation-Space(X, $BDO_{X+Y}(k)$) = {¬∃s (placing(X, mark(X), s, k): $B_{X(k)}$¬ turn(X, k)}. This can be proved based on (Ap.1.i2) and (OB3).

(e3) Permission-Space(X, BDO_{X+Y}(k)) = {placing(X, mark(X), s, k): $P_{X(k)}$ placing(X, mark(X), s, k)}.

f. Schema for dynamic development of Tic-Tac-Toe

(f1) [Action in a continuing stage] In case: $B_{X+Y(k)}\neg$ end(k) & $B_{X+Y(k)}$ (turn(X, k) ∧ opponent(X, Y)). From (Ap.4.e) follows: Obligation-Space(X, k) = {$\exists^{=1}$s (placing(X, mark(X), s, k) ∧ vacant(s, k))} & Obligation-Space(Y, k) = {¬∃s (placing(Y, mark(Y), s, k)} & Permission-Space(X, k) = PLACING(X, mark(X), List(k), k) & Permission-Space(Y, k) = Ø. We assume here that X desires to place mark(X) in position s_1. Thus, DB_X(k) = DB_X(k - 2) ∪ {placing(X, mark(X), s_1, k)}. As special cases, we define: $DB_A(-2) = DB_A$ & $DB_B(-1) = DB_B$. Then, X decides to perform this action and both X and Y confirm that this action is performed. Thus, we set: BB_{X+Y}(k + 1) = BB_{X+Y}(k) ∪ {placing(X, mark(X), s_1, k)} & List(k + 1) = List(k) – {s_1}. Then, based on (ET6), we can infer: $B_{X+Y(k+1)}$ (occupied(s_1, mark(X), k) ∧ turn(Y, k + 1)).
(f2) [Action in the terminal stage] In case: $B_{X+Y(k)}$ end(k). The game ends here. Because of (OB4), it holds: Permission-Space(X, k) = Ø & Permission-Space(Y, k) = Ø.

g. [Translation into Propositional Logic] Each FO-sentence in BBttt and in OBttt is reducible to a propositional sentence, because the ranges of all variables in them are finite. For example, ∀x φ(x) can be translated into $\varphi(a_1) \wedge \ldots \wedge \varphi(a_k)$ and ∃x φ(x) can be translared into $\varphi(a_1) \vee \ldots \vee \varphi(a_k)$, where $a_1, \ldots, a_k$ are names for objects whose set is a range of variable x.

(Ap.5) [Description of game developments] The following sequence of formulas describes the game development discussed in Subsection 3.1: ⟨placing(A, O, [1,1], 0), placing(B, ×, [2,2], 1), placing(A, O, [3,3], 2), placing(B, ×, [1,3], 3), placing(A, O, [3,1], 4), placing(B, ×, [3,2], 5), placing(A, O, [2,1], 6)⟩. I call this sequence of actions "the record of a game". In general, a game of Tic-Tac-Toe can be described by a structure in form: ⟨⟨$BB_G(0), OB_G$⟩, ⟨$BB_G(1), OB_G$⟩, … , ⟨$BB_G(k), OB_G$⟩⟩. I call this structure "the game history". In general, this game history can be constructed for any game that is describable by Dynamic BDO-Logic. For example, a game history can be constructed for a baseball game and a soccer game.

(Ap.6) [Role-specific normative spaces] For any agent A in group G, role-specific obligation, prohibition, and permission space are characterized as follows (here, we assume that $\varphi_{act}(A)$ is a sentence, which expresses that A performs certain action).

a. Obligation-Space*(A, $role_1$, G, t, ⟨BB_{NS}, OB_{NS}⟩) = {$\varphi_{act}(A)$: $O_{NS}\forall$ x ($role_1$(x, G, t) → $\varphi_{act}(x)$) & $B_{A(t)}$ $role_1$(A, G, t) & $O_{A(t)}$ $\varphi_{act}(A)$}.

b. Prohibition-Space*(A, not($role_1$), G, t, ⟨BB_{NS}, OB_{NS}⟩) = {$\varphi_{act}(A)$: $F_{NS}\exists$ x (¬$role_1$(x, G, t)∧ $\varphi_{act}(x)$) & $B_{A(t)}\neg$ $role_1$(A, G, t) & $F_{A(t)}$ $\varphi_{act}(A)$}.

c. Permission-Space*(A, $role_1$, G, t, ⟨BB_{NS}, OB_{NS}⟩) = {$\varphi_{act}(A)$: $P_{NS}\forall$ x ($role_1$(x, G, t) → $\varphi_{act}(x)$) & $F_{NS}\exists$ x (¬ $role_1$(x, G, t)∧ $\varphi_{act}(x)$) & $B_{A(t)}$ $role_1$(A, G, t) & $P_{A(t)}$ $\varphi_{act}(A)$}.

Author details

Yasuo Nakayama
Graduate School of Human Sciences, Osaka University, Osaka, Japan

*Address all correspondence to: nakayama@hus.osaka-u.ac.jp

References

[1] Davidson D. Essays on Actions and Events. Oxford, UK: Oxford University Press; 1980

[2] Bratman M. Intention, Plans, and Practical Reason. Cambridge, MA: Harvard University Press; 1987

[3] Searle JR. Making the Social World: The Structure of Human Civilization. Oxford, UK: Oxford University Press; 2010

[4] van Bentham J. Logical Dynamics of Information and Interaction. Cambridge: Cambridge University Press; 2011

[5] Baltag A, Renne B. Dynamic epistemic logic. In: Zalta EN, editor. The Stanford Encyclopedia of Philosophy. Stanford: Stanford University; 2016

[6] Gärdenfors P. Knowledge in Flux: Modelling the Dynamics of Epistemic States. Cambridge, MA: MIT Press; 1988

[7] Hansson SO. Logic of belief revision. In: Zalta EN, editor. The Stanford Encyclopedia of Philosophy. Stanford: Stanford University; 2022

[8] Kamp H, Reyle U. From Discourse to Logic. Dordrecht: Kluwer; 1993

[9] Geurts B, Beaver DI, Maier E. Discourse representation theory. In: Zalta EN, editor. The Stanford Encyclopedia of Philosophy. 2020

[10] Jankovic M, Ludwig K, editors. The Routledge Handbook of Collective Intentionality. London: Routledge; 2018

[11] Nakayama Y. Norms and games as integrating components of social organizations. In: Kasaki M, Ishiguro H, Asada M, Osaka M, Fujikado T, editors. Cognitive Neuroscience Robotics B: Analytic Approaches to Human Understanding. Tokyo: Springer Japan; 2016. pp. 253-271

[12] Nakayama Y. A formal analysis of legal reasoning. In: Proceedings of the Tenth International Workshop on Juris-Informatics (JURISIN 2016); 14–15 November 2016; Kanagawa. Tokyo: JSAI; 2016. pp. 3-16

[13] Nakayama Y. Philosophical basis for dynamic belief-desire-obligation logic. In: Kyoto Philosophical Logic Workshop III. Kyoto; 2017

[14] Nakayama Y. Agents, actions, and social reality. In: Proceedings of SOCREAL2022: 6th International Workshop on Philosophy and Logic of Social Reality. Sapporo: Hokkaido University; 2022

[15] Bratman M. Two faces of intention. In: Mele AR, editor. The Philosophy of Action. Oxford, UK: Oxford University Press; 1997. pp. 178-203

[16] Searle JR. Intentionality: An Essay in the Philosophy of Mind. Cambridge: Cambridge University Press; 1983

[17] Searle JR. Speech Acts: An Essay in the Philosophy of Language. Cambridge: Cambridge University Press; 1969

[18] Searle JR. Expression and Meaning: Studies in the Theory of Speech Acts. Cambridge: Cambridge University Press; 1979

[19] Bratman ME. Shared Agency: A Planning Theory of Acting Together. Oxford: Oxford University Press; 2014

[20] Tuomela R. Social Ontology: Collective Intentionality and Group Agents. New York: Oxford University Press; 2013

[21] Ludwig K. From Plural to Institutional Agency, Collective Action: Volume 2. Oxford: Oxford University Press; 2017

[22] Weber M. Soziologische Grundbegriffe. In: Weber M, editor. Wirtschaft und Gesellscaft. Tübingen: Mohr; 1922

[23] Durkheim É. De la Division du Travail Social. Paris: P.U.F; 1893

[24] Lewis D. Convention: A Philosophical Study. Cambridge, MA: Harvard University Press; 1969

Chapter 5

Employing Games of Partial Information for Understanding Microaggressive Conflicts

Corey Reutlinger

Abstract

The term microaggression is used to describe everyday ambiguous slights or "put downs" that communicate discrimination toward a person belonging to a marginalized group. Longstanding critiques of the term have included unclear conceptual boundaries, forced casual linkages to mental health effects, and inadequate contextual criteria for identifying their occurrences. Recent research suggests that employing linguistic principles such as syntax, semantics, and pragmatics can help build an interactive examination for the study of conditions that influence a microaggressive encounter. Situated games of partial information are necessary for a computationally-tractable analysis of the textual, contextual, and interdependent features of ambiguous communicative exchanges. Thus, this chapter describes a microaggressive exchange between communicators by using situated games of partial information. Specifically, I detail a conversation excerpt where a microaggression emerged during a social interaction. I show how a communicator can interpret a message as discriminatory by examining the syntactic, semantic, and pragmatic factors of the microaggression through games of partial information. This approach invites communicators to account for the many probabilistic conditions that inform a microaggressive exchange so that they can begin to repair discriminatory comments without defaulting to prescriptive responses that potentially escalate social hostility.

Keywords: microaggressions, games of partial information, pragmatics, organizational conflict, disability

1. Introduction

"You're such an inspiration!"; "Aren't you being oversensitive?"; "You're crazy!" Perhaps you have experienced a situation where you may have received one of these well-meaning "put downs." Nearly 50 years of research in the field of psychology has attempted to capture how groups of people experience these everyday, ambiguous slights known as *microaggressions*. These subtle, automatic, often unconscious messages communicate denigration toward people who identify with and belong to marginalized groups [1, 2]. While well-intentioned to mean "no harm done," any person who communicates a microaggression to a receiver is often unaware that what is said or done can produce distressing effects in the receiver's mind and body. Such messages illuminate dynamic tensions of systemic oppression such as racism, sexism, heterosexism, ableism, etc., and how such structures influence the way people behave in their social interactions [3, 4]. In recent years, communication researchers have veered away from psychology to data-driven analyses of linguistic principles such as syntax, semantics, and pragmatics to build interactive examinations for the study of conditions that influence a microaggressive encounter [5–8].

This chapter attends to one type of pragmatic analysis for the study of microaggressions using game theory as a guiding tool. It describes a microaggressive exchange between communicators by using situated games of partial information. Specifically, I detail a conversation excerpt where a disability microaggression emerged during a social interaction—a relatively understudied type of microaggression in current works [9]. First, I outline a few assumptions of situated games of partial information as it applies to microaggressions. Next, I describe a brief encounter of a naturally occurring disability microaggression in an organizational setting. Finally, I analyze what may have contributed to this encounter. I show how a communicator can interpret a message as discriminatory by attending to the syntactic, semantic, and pragmatic factors of the microaggression. This approach invites communicators to account for the many probabilistic conditions that inform a microaggressive exchange so that they can begin to repair discriminatory comments without defaulting to prescriptive responses that potentially escalate social hostility [7, 10].

2. Microaggressions and game theory

Most criticisms of microaggression research have called these subtle snubs both imaginary and trivial. In fact, longstanding critiques of the term have included unclear conceptual boundaries, forced casual linkages to mental health effects, and inadequate contextual criteria for identifying their occurrences [11]. A lack of clear operationalization in psychological experimentation to delineate specific message structures as microaggressive has prevented this research area from expanding; this is because biases, varying degrees of consciousness, and speaker intentions are not easily operationalizable to control variables [12]. Communication research has attempted analyses of pragmatics—e.g., the grammars, tones, contexts, turn-taking, etc. [13]—to resolve such critiques [5, 8]. Nonetheless, the scarcity of contributions to this niche research area has beseeched scholars to contemplate new methodological approaches to fully capture the factors that influence the prejudicial nature of these inexplicit, ambiguous, and harmful messages.

One way to attend to the pragmatics of microaggressive communication patterns in social interactions is through game theory. *Game theory*, or multi-person choice theory, is the mathematical study of strategic interaction between independent, self-interested individuals [14]. Thus, this particular mathematics is equipped to explain how a communicator's choice of a social action is influenced by the actions of other communicators during a situation (and is also influenced by the hidden markers of biases, prejudices, intentions, and degrees of consciousness in a communicator). Communication scholars first restricted this mathematics to the study of interpersonal conflicts [15]. However, game theory is valuable for assessing all types of complex communicative processes, regardless of the situation, including nuanced encounters such as microaggressive ones. It offers a vehicle for understanding how everyday communication can become power-laden, harmful, discomforting, and silencing for some persons who identify along marginalized societal lines.

Most mathematical models have historically aimed to describe communication dynamics in social interactions disjointly, focusing more so on context-free approaches to the study of social interactions [16]. Game theory, however, attends to the interdependent ways microaggressions act to draw on text and context to invalidate personal cultural identities and sociopolitical experiences [6, 7, 17]. *Text* is the tangible sentences that obtain messages, which act as well-behaved, rule-governed objects for conveying a communicator's intentions and information [18]. *Context* includes the (often unspecifiable) circumstances that regulate what meanings emerge and attach to messages, which includes referents, symbols, discourses, histories, and smaller sequences of conversations [17, 19]. The situatedness in communication presents a unique scientific challenge because microaggressions thrive off of incomplete, ambiguous information between senders and receivers [12]. Game theory offers a context-sensitive approach and computationally-tractable analysis of the textual, contextual, and indeterminate features of these problematic communicative exchanges [17, 19]. In other words, it can highlight what conditions may produce and influence the discomforting "content" during microaggressive conflicts if we were to examine how ambiguity dictates the social actions for both the sender and receiver in these situations. I offer a way of examining microaggressions, specifically, through situated games of partial information [7, 17–19].

2.1 Microaggressions as situated games of partial information

Game theory treats social interactions as *games*, or well-defined mathematical objects consisting of distinct characteristics [14]. I define a game as a type of communicative activity that includes the following six elements [20]:

1. *Communicators*: The number of rational-thinking individuals present during an interaction as well as their respective turns.

2. *Actions*: What do communicators do when they interact with each other? How long do they talk? What questions should they ask? What words do they say?

3. *Situations*: Communicators enter an interaction believing different assumptions about what's going on. This includes what (and how many) actions communicators think they can use to interact with each other. This can vary per communicator.

4. *Initial probabilities*: Communicators subconsciously assess the effectiveness of their choices before making them. They weigh what kinds of meanings their words and messages have before expressing them. Each communicative choice varies in how well it will be received and interpreted.

5. *Information sets*: What kinds of knowledge do communicators bring to the situation? This can be both personal and shared knowledge about contexts. Communicators use what they know to decide what choices to make. This can range from their personal histories, cultural understandings, or what social rules they use when interacting.

6. *Payoffs*: These are the outcomes of a situation. A communicator may interpret a word, message, or action in several ways. This all depends on what a person's underlying preference is during an interaction. Do they want to cooperate? Do they want to have conflict? What would be the costs or benefits?

These six elements must be accounted for in some manner when analyzing microaggressive communication through games.

There are also two unique pragmatic properties that microaggressions obtain, which complicate comprehension and cooperation during a social interaction. The first property is irrationality. One assumption in game theory is that communicators interact with each other through logical thoughts and behaviors. Unfortunately, communicators are sometimes *irrational*, and behave in a way that reflects their hidden biases, heightened emotions, and unconscious intentions [17]. People may not know why they say or do the things they communicate during an interaction. The second property is indeterminacy, or uncertainty. Most people assume the other party will cooperate. The presence of *indeterminacy* means a person's intentions can never be known, social rules people follow when interacting are never explicit, and communicators perceive what is happening in a situation uniquely. Both properties make it possible for microaggressive encounters to be noncooperative and unresolvable.

Some linguists suggest microaggressions likely equate to Bayesian game classification problems or strategic games of incomplete information [6, 17]. Under this assumption, communicators do not share a common understanding of the situation between them—yet, they do share the same kinds of communication strategies and social rules on how to interact with each other. As previously mentioned, this is not always the case, especially when irrationality and indeterminacy are in the mix. Intercultural nuances in language imply that people do not always operate with the same social rules or ways of communicating; people misunderstand each other, make assumptions about each other's identities, or are vague when crafting messages. Thus, people interact more from a space of *partial* information: they communicate with various types of strategies, follow different social rules, and draw on multiple types of background knowledge and biases to inform their choices in a social interaction, sometimes even hidden from their own personal awareness [17, 19]. The more appropriate way to conceptualize microaggressions is through *situated games of partial information*, or games of noncooperation where communicators draw from varying degrees of personal and contextual knowledge. I outline the framework necessary to examine a microaggression encounter through a game of partial information (for a deeper reading, construction, and outline of the framework, see [17, 19]).

2.2 The framework

Understanding how and when microaggressions occur is a complex and unpredictable process. Examinations of microaggressive encounters must include the social interaction, the context, and the larger discourses that shape what is happening between communicators. Microaggressive social interactions as situated games occur on two levels. First, they occur through the little-d, microdiscourses of conversation, or the particulars of talk and text. Second, they occur through the big-D, macrodiscourses, or the larger institutional and cultural practices, the sociopolitical ideologies, and historical narratives that influence and are influenced by conversations communicators have with each other [7, 19]. These two levels create a circular process for conveying and interpreting messages between communicators [19]. There are four large interlocking parts (or steps) needed for dissecting a microaggressive event, according to reference [19]: the (a) Setting Game, (b) Content Selection Game, (c) Generation or Encoding Game, and (d) Interpretation or Decoding Game. The overall communication process for playing situated games is shown in **Figure 1**. Each number indicates a "step" in the process between communicators.

The first game is the Setting Game. Communicators play an unconscious game that attends to cultural, historical, and political ideologies that guide social rules for interacting. **Step 1.** This game starts when some situation or circumstance compels communicators to interact with each other. **Step 2.** The setting then triggers communicators either (a) to send messages so that ideas can be exchanged or (b) to elicit an action or response. The Setting Game acts as a decision problem of some kind [19]. It dictates what kinds of obvious (or hidden) goals and subgoals communicators have and whether they want to cooperate with each other. The Setting Game is nebulous and therefore is better as a descriptive account for analyzers of microaggressive situations. Analyzers should describe what constrains and explains the interplay between communicators and what their future actions might be like. Microaggressive encounters can begin the moment communicators enter a similar environment and use that environment to interact with each other.

Second, once communicators decide to send messages or elicit responses from each other, they enter a Content Selection Game. **Step 3**. Communicators tap into their information sets to draw from an assortment of vocabularies, languages, biases, ideas, discourses, histories, presuppositions, social rules, and so on. **Step 4.** Then, they play a quick, thinking game to decide what combinations of words will best craft a message that fulfills their intentions (or rather, optimizes their content). Of course, this game is tricky because communicators do not always know how someone will respond to a message until its expressed [19]. Further, biases and emotions can complicate what words to choose (i.e., messages always sound better in the mind until they are uttered). This game is about assessing what personal knowledge communicators use to engage with each other. Because analyzers do not know what communicators are

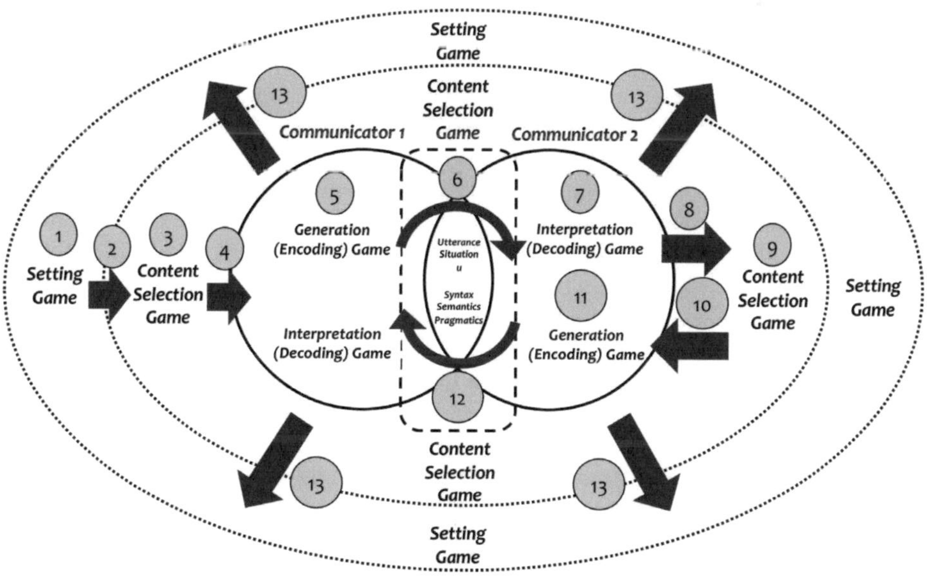

Figure 1.
Communication process for playing situated games.

thinking (and sometimes even communicators do not know what their own thoughts are about), this game can only be speculated through uttered conversation. Analyzers should segment full conversations into shorter parts, providing enough of the conversation to scrutinize the microaggressive encounter and to contextualize the setting.

Third, once a communicator selects the best content they want to convey from a setting, they enter the Generation or Encoding Game. **Step 5.** Communicators think about (and sort through) possible alternatives to the message they want to express; they imagine how someone might respond to and interpret their alternatives [19]. Of course, communicators narrow down their choices to one message because they want to express the best one. **Step 6**. Based on the various types of syntactic, semantic, and pragmatic factors that can optimize their choice in the current utterance situation, called *u* (as indicated by the dotted-line box in the center of **Figure 1**), communicators then use a shared language, like English or Spanish, to express their chosen message [17, 19]. Most analyses of games typically focus on the messages uttered; however, the Encoding Game is unique in that it considers alternative messages that were crafted, but not uttered. In other words, communicators who encode and send messages have multiple starting points (rather than only one). Thinking about alternative messages may stem from explicit and implicit biases for a communicator; these alternatives dictate whether a receiver of a message will cooperate or not with the sender. Thus, communicators are (sometimes) careful to construct and send their messages during a social interaction so it does not lead to noncooperation like that in a microaggressive conflict. Analyzers should account for the syntax, semantics, and pragmatics of the messages uttered in a conversation segment; additionally, analyzers should consider possible alternative messages communicators may have thought to use.

Finally, a communicator who receives a message will play the Interpretation or Decoding Game. **Step 7**. This game can happen either sequentially or simultaneously with the Encoding Game. Essentially, once a sender expresses their chosen message, it becomes the receiver's turn. **Step 8**. Receivers must infer what content is being conveyed. **Step 9.** So, they return to their own Content Selection Game to infer what other information is influencing the situation in order to craft their own best responses [19]. Most game-theoretic analyses account for how receivers interpret a sender's message when there is complete information available to them. Because the receiver in a microaggressive situation does not know a sender's intentions or biases when a message is communicated, analysis of the Decoding Game must account for how a receiver interprets incomplete or partial information. That is, it must consider how receivers make sense of a message based off alternative messages that could have been expressed by the sender to communicate a hidden intention. **Step 10.** After thinking through and selecting content to infer an interpretation, receivers decide what combinations of words can be used to best craft a response. **Step 11.** Receivers will start their own Generation or Encoding Game, thinking through alternative ways to frame their response. **Step 12.** At last, they select the best response based on what was inferred by the original message. **Step 13.** Both communicators return to the Setting Game and repeat the process in a back-and-forth, turn-taking style until the interaction ends.

These four interlocking games build a structure for analyzing the particulars of talk while regarding partial information that stems from hidden biases, ideologies, histories, and contexts. Data from naturally occurring conversations reify these conceptual games into tangible ones to be used for a precise examination of the many pragmatic conditions that contribute to a microaggressive encounter. The next section describes a brief, recorded, conversation segment of a microaggressive social interaction I had with an American Red Cross (ARC) representative to discuss an organizational issue during the height of COVID-19 pandemic.

3. The case of the American red cross

I employ situated games of partial information for examining how a disability microaggression unfurled from a specific, 50-minute, communicative exchange I had with an ARC representative. I briefly discuss the Setting Game and a specific excerpt that is appropriate for understanding the Content Selection Game. I end by analyzing the turn-taking between the Generation and Interpretation Game to understand how communicators generally label a message as a microaggression.

3.1 The setting game

The American Red Cross is a non-profit humanitarian organization that offers several opportunities for education and training in emergency assistance, disaster relief, and disaster preparedness. One example of this would be providing education and certification for safety in aquatic environments such as pools, beachfronts, water parks, and so on. The American Red Cross prides itself as being one of the main organizations for certifying everyday patrons to become lifeguards across many regions in the United States. It also prides itself in providing training and certification for individuals to become Water Safety Instructors, or teachers who can demonstrate to young children, teens, and adults how to swim proficiently and safely in aquatic environments. Every 2 years, ARC instructors in the United States must recertify their credentials by teaching courses and/or attending in-person review assessments of their teaching skills.

During the 2020 COVID-19 pandemic, the ARC updated their main website and their overall record-keeping system. The transition to the new management system meant that, for several instructors across the United States, their digital records were either misplaced, expunged, or marked as "expired." At one point, I had noticed my Water Safety Instructor certification was no longer active in the digital system (i.e., it was labeled as expired). Even though I had completed annual recertifications to maintain my credentials, the new record-keeping management system should have some information about the online whereabouts of my credentials such that it would display as active and not as expired.

Confused, I called an ARC representative to discuss the issue after a string of email exchanges with a customer service representative failed to resolve the issue. In other words, the setting compelled us to interact with each other through conversation to solve the problem at hand. Both of us entered this space to resolve the conflict, aiming to understand and problem-solve how the certificate had been marked as expired. Both of us were knowledgeable about the ARC as an organization and its policies, the recent transition that occurred to update the overall system, and what barriers had emerged in the organization and for instructors across the United States because of the pandemic. Our goals and subgoals during the 50-minute telephone conversation were to engage in cooperation while resolving the issue pertaining to credentials.

3.2 The content selection game: an excerpt

Once I connected with the ARC representative, we began to discuss the problem over the telephone. The following is a brief excerpt of the conversation leading up to what I perceived to be a disability microaggression. While the microaggression is part of both the Encoding and Decoding Games, the conversation informs the types of turn-taking, assumptions, and possible emotions that contributed to this game. It goes as follows:

[18:13] ARC Rep: OK. .... here's what I've looked up. I went ahead and looked up your information that we have in the system.

[18:43] Me: OK.

[18:44] ARC Rep: Your Water Safety Instructor certification was expired, correct?

[18:50] Me: Well, it should not be expired. That's the thing. It was showing that it was.

[18:55] ARC Rep: OK, it's showing it was expired in 2018 .... So, if you have any proof—which I believe you—if you have any proof at all ... I can go in and get this fixed .... But I will need to have proof that shows that you recertified in 2018 because, as of what our system shows, it says that you expired.

[20:33] Me: [pause] ... Yeah.

For most readers, this may seem like a regular exchange between two communicators working to solve a technological issue. However, for some readers, this segment aligns with a microaggressive moment (as was the case for me). The use of the word "proof" by the ARC rep follows one of eight themes from a recognized disability microaggression taxonomy: denial of a disability-related experience [9]. In the study that developed the taxonomy, one focus group interviewee had stated, "Because I don't have an outward disability, people don't necessarily believe me. I've had to deal with that all my life, and I've had to give proof" [9]. This also follows a "do not look sick" cultural assumption that people with invisible disabilities must prove through documentation that their bodily experiences (around disability) are valid and not imaginary [21]. Even when a person's disability is not the central contextual focus of a conversation, such as in the excerpt above, using the word "proof" when communicating a message can function in a similar fashion to invalidate a person's lived experience or perceptual reality (despite being told "I believe you"). It is an effort by perpetrators (whether they meant to use it in a denigrating way or not) to deny responsibility for any difficulties an inaccessible barrier might create [9]. Receivers can experience emotions of discouragement or confusion; the turn-taking here shows an imbalance between how much each communicator says. Thus, ableist assumptions, power differentials, and mismatched emotions are some of the potential pragmatic factors that inform how the communicators might have selected the content for their messages. Encoding and Decoding Games help to specify these contents' whereabouts (and make sense of how a word can trigger a microaggression for some people while not for others).

3.3 The generation (encoding) game

The Generation or Encoding Game is about turning speaker intentions into an expressed message. This also classifies the syntax, semantics, and pragmatics of a message and considers the likelihood a message carries a sender's intentions accurately across what is being conveyed.

3.3.1 The syntax

First, we need turn to the syntax of a message, or the parts of speech. If we were to focus on "proof" as the triggering word, we notice that "proof" is being used as a noun in the message "I will need to have proof" and "If you have any proof at all" (I will focus on the first example for simplicity). In Ref. [17, 19], the algebraic system $(T, \bullet)$ is used to describe a context-free grammar G as a system of syntax parse trees (or string of words, phrases, and whole sentences) with a product operation. So, let t symbolize a syntactic parse of a word (e.g., "proof" functions as a noun). Also, let • be an operator that "chains" together the syntactic parses of words into a full syntax parse tree for a sentence S (i.e., it "multiplies" parses of words together to form a string, such as $t_i \bullet t_j = t_i t_j$ where i and j are indexes for the order of a word in a sentence) [19]. In the sentence "I will need to have proof," the word "proof" has the following parse in Eq. (1):

$$t_6 = [_{Noun}\ \text{PROOF}] \tag{1}$$

Because "proof" is the sixth word in the sentence, and only functions as a noun, it has one syntactic parse. Every other word in the sentence also has its own unique parse and function in the full sentence of the message. Together, each word's parse becomes $t_1 t_2 t_3 t_4 t_5 t_6 = \varphi$, where φ denotes the entire expressed utterance, "I will need to have proof," which stems from a natural language $\mathcal{L}$. The full utterance can also be represented uniquely as a string of words such as $\varphi_1 \circ \varphi_2 \circ \varphi_3 \circ \varphi_4 \circ \varphi_5 \circ \varphi_6$, where $\circ$ denotes a "grammatical concatenation" operation used to generate the content of the utterance, "I will need to have proof" (this means $(\mathcal{L}, \circ)$ is an algebraic system to describe a given language) [19]. Therefore, "proof" can also be represented as follows in Eq. (2):

$$\varphi_6 = \text{PROOF} \rightarrow [_{Noun}\ \text{PROOF}] = t_6 \tag{2}$$

The symbol $\rightarrow$ acts as a *syntactic map* for transforming words into their possible speech parses. Classifying syntax narrows down the many possible denotations and connotations a word could mean ("proof" can sometimes be used as an adjective or verb), making game analysis an easier process.

3.3.2 The semantics

Second, we turn to semantics, or the dictionary definitions and contextual referents for the noun, "proof." In this example, "proof" has one dictionary meaning that fits: "evidence that establishes a fact or the truth about a statement." Also in this example, "proof" has at least two contextual meanings—one that means "certification" and the other "verification." Both are distinct because the first refers to the ARC Water Safety Instructor certificate and the other refers to verifying a perspective, a lived experience, or a way of understanding what is happening in a conversation. Both are needed to understand intention.

Formally, the word "proof" is first mapped into its dictionary or conventional meaning (whether that be a property, P, or a relation, R, of language $\mathcal{L}$) with a *conventional map*, denoted by $\rightarrow$ [17, 19]. The conventional meaning of the word "proof" is then mapped into its two contextual or referential meanings relative to some contextually-bounded utterance situation called u via a *referential map*, denoted by $\rightarrow_u$ [17, 19]. So, "proof" can be represented semantically as follows:

$$1.\text{Referential Use}: \varphi_6 \rightarrow P^{\varphi 6} \rightarrow {}_u\sigma_6.$$
$$2.\text{Referential Use}: \varphi_6 \rightarrow P^{\varphi 6} \rightarrow {}_u\sigma_6'.$$

The symbols, σ_6 and σ_6', denote the possible semantic values for φ_6 = PROOF in the current setting (i.e., obtaining a property that either means "certification" or "verification"). Every other word also has their own unique conventional and referential meanings. Together, these meanings create the possible "contents" of an utterance given the situation where the utterance is expressed.

3.3.3 The pragmatics

Third is the pragmatics, or what is being implied over what is said and done. Because there are at least two ways the word could be interpreted, "proof," when used in an utterance, could either imply (a) (→) [*We need a certificate to match our records*] or (b) (→) *[We need to validate that you are not faking your experience]*.

In mathematical notation, the word "proof" can obtain two pragmatic issues that need resolving. They are as follows in Eq. (3) and Eq. (4):

1. Implied Meaning (a) with respect to pragmatic issue:

$$\varphi_7 = \varphi = \varphi_1\varphi_2\varphi_3\varphi_4\varphi_5\varphi_6 \rightarrow {}_u\sigma_1\sigma_2\sigma_3\sigma_4\sigma_5\sigma_6 \rightarrow {}_u\sigma_7 \quad (3)$$

2. Implied Meaning (b) with respect to pragmatic issue:

$$\varphi_7 = \varphi = \varphi_1\varphi_2\varphi_3\varphi_4\varphi_5\varphi_6 \rightarrow {}_u\sigma_1\sigma_2\sigma_3\sigma_4\sigma_5\sigma_6 \rightarrow {}_u\sigma_7' \quad (4)$$

The symbols, σ_7 and σ_7', denote how the utterance "I will need to have proof" transforms its syntactic and semantic contents into the two implied meanings (→) [*We need a certificate to match our records*] and *[We need to validate that you are not faking your experience]*, which aim to resolve the pragmatic issues of communication happening in this utterance situation *u* [19]. While these two implied meanings are triggered by the word "proof," they are fully supported by the entire utterance because it requires all the words of the utterance to imply meaning beyond what is being stated.

While the ARC rep clarified their intention, the word "proof" can still be triggering for someone (like myself) who has been told countless times to prove the existence of their body, their disabilities, and their experiences. Microaggressions hold a unique property of being cumulative over many unrelated interactions [1, 2, 4]. In other words, receiving the word "proof" to mean "verification of one's own lived experience" countless times across different contexts and situations can acquire a denigrating meaning for persons with disabilities (even during times when the word "proof" is not being used in a situation to mean "verification").

The classification of syntax, semantics, and pragmatics for the word "proof" outlined in this section is a condensed version of a full Encoding Game. In actuality, communicators play an Encoding Game for each word, phrase, and utterance being expressed in a conversation. I have chosen to focus efforts on the word "proof" for the purpose of outlining the two implied meanings, which are necessary for the Decoding Game. It is possible to see how a communicator can decode a message and label it as a microaggression when partial information exists inside a situation.

3.4 The interpretation (decoding) game

The Interpretation or Decoding Game disambiguates (or clarifies) the content from a message's syntactic, semantic, and pragmatic possibilities [19]. It accounts for how communicators interact: how they send, interpret, and respond to messages while text, context, irrationality, and indeterminacy are interwoven into and evolving throughout the social interaction.

Define G to be a game of possible social actions between communicators in the current utterance situation u. Define g as a mathematical function of G that transforms the words, phrases, and implied meanings of an utterance into a game (g is a container for syntax (t), the text (φ), and context (σ)) [17, 19]. As a receiver, I can decide whether I want to cooperate with the ARC rep and interpret "proof" as either the implied meaning (a) or (b) from above. I could also choose to not cooperate and be conflictual (meaning, I could misunderstand his message and could respond in a way that escalates the tension and power differentials between us). Focusing on the implied meanings, we have the following in Eq. (5):

$$g_7(\text{Implied Meaning of Utterance}) = g_7 \tag{5}$$

The function, g_7, transforms the implied meanings into an algebraic game that aims to resolve the pragmatic issues discussed earlier, and does so through a game, or specifically through "game 7." Using game transformations adds content such as pragmatics to create a context-sensitive grammar. These functions also stem from the algebraic system (G, $\otimes$), where the symbol $\otimes$ denotes a game multiplication operation that multiplies together the many possible syntactic, lexical, phrasal, sentential, semantic, and pragmatic games of an utterance [17, 19]. The games act as "go-betweens" for understanding how receivers interpret and respond to a sender's message(s). Because microaggressions stem primarily from a discernment (and frequent disagreement) of implied meanings in a message, I will focus this Interpretation Game to game 7 for simplicity. The overall game is shown in **Figure 2**.

The top half represents the real situation (what happened). The game begins with the ARC rep, noted by a square. The square is labeled with the type of initial situation, s_{real}, and the probability, p_{real}, or the chance the ARC rep conveys "certification" as their intention when saying "proof" in an utterance. It's then my turn as the receiver. This is noted by the circle, r_{real}, in the responding situation[1]. There are two possible outcomes. Either I cooperate and interpret the message as (a) because I think "proof" means "certification." Or, I decide not to cooperate and instead interpret the message as (b), leading to conflict. Clearly the payoff[2] A (where we both benefit) is better than B (where we both could end up with a cost) because I want my interpretation to align with the ARC rep's intention.

However, let us assume the opposite. The bottom half represents an alternative situation, one where the ARC rep may have some hidden motives (even unknown to themselves) for using the word "proof." The game would begin again with the ARC

---

[1] The oval enclosing both top and bottom half represents the information set, or the amount of knowledge I have available to interpret a message correctly [19]. To account for a speaker's intentions, which is something I do not know, I must consider an alternative situation. Without the oval, I could make different choices in either half of the game tree, which restricts the possible solutions of the game [19].

[2] Most payoffs are written in the general form of (a_A, a_B), (b_A, b_B), and so on. Payoffs can also have numerical values such as (4, 2), (7, −5), and so on. These are arbitrary at best [17–19].

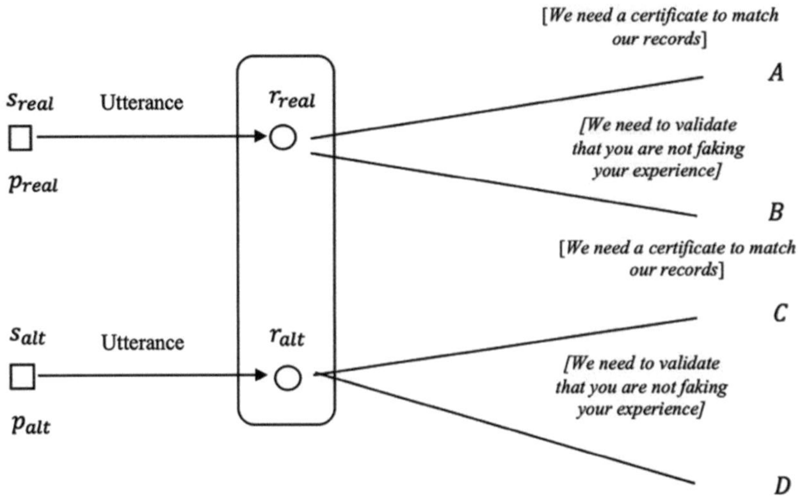

Figure 2.
Game resolving the pragmatic issue in utterance.

rep in the alternative situation, s_{alt}, with probability[3], p_{alt}, or the chance the ARC rep conveys "verification" as their intention when saying "proof." When it is my turn (represented by r_{alt}), I once again have two outcomes to choose from, but they are treated differently because the intention this time might be more ill-natured. Either I interpret the message as (a), which means I'm not cooperating (and I'm lying to myself, thinking the ARC rep is a good person that did not micro-aggress me). Or, I interpret the message as (b) because I think "proof" means "verification." If I choose (b), I cooperate, which could spark further conflict because I know that the ARC rep was intending to be ill-natured when sending their message. This may also explain my pause and response of "... Yeah" after the ARC rep spoke. Those who recognize themselves in a lower status or minoritized position of power may hesitate to respond (or stay silent) when conversing with someone of a higher status who expresses a problematic or troublesome utterance [7, 10]. Although seemingly counterintuitive, D is better than C because I want us to cooperate (even if I know a heated discussion is possible afterwards).

This game shows how an implied meaning from an utterance can be interpreted in a way to arrive at a disability microaggression (e.g., represented by the payoffs B or D). My response of "... Yeah," highlights my own hesitancy in processing what I received as a message from the ARC rep. In the moment, the ambiguity heightened power asymmetries, emotional confusion, and communicative hesitancies. I discover later in the same conversation the ARC rep used other word combinations that suggested an ableist bias hidden in their intention. In which case, the outcome C in the alternative situation could have been the actual best payoff for us both (i.e., because he holds power, I must interpret a message as if it did not have an ableist bias or logic embedded into its content—otherwise I risk escalating conflict or a quick end to our conversation).

---

[3] Both p_{real} and p_{alt} act as functions of what conditioning variables (such as syntactic parses and semantic values) influence the utterance situation u. Therefore, the speaker conveys content via a probability distribution (p_{real}, p_{alt}) when choosing an utterance to express. This also means $p_{real} + p_{alt} = 1$. For more information, see [17, 19].

4. Conclusion

This chapter offered a shortened, pragmatic analysis of microaggressions using game theory as a guiding tool. I described how situated games of partial information [17, 19] can be used to understand how microaggressive messages obtain their syntactic, semantic, and pragmatic contents during social interactions. In other words, these types of games are helpful for classifying and explaining how senders and receivers of communication make decisions when given ambiguous conditioning variables. It requires an understanding of the setting, the way communicators select content such as ideologies, discourses, referents, etc. from that setting, and encode and decode those contents when exchanging messages through turn-taking [19]. I detailed a conversation excerpt between myself and an American Red Cross representative, which led to the emergence of a potential disability microaggression during our social interaction. Partial information games offer a holistic approach for recognizing how unknown speaker intentions can be mismatched with a receiver's interpretations during ambiguous moments of communication [7, 17–19].

To be clear, this particular chapter covers a simplified version of the complex dynamics in a conversation. Essentially, it covers only one game of the many that are being played simultaneously, sequentially, and nonlinearly on a syntactic, semantic, and pragmatic level of interaction. The reader should consider a few things. First is calculating the chance that the ARC rep is being well-meaning or ill-natured with their intentions. The conversation excerpt provides some context for the turn-taking; however, most conversations do not offer an exactness of a sender's intention nor a reason why a receiver responds in the way they do. Situated games of partial information can also be applied to the whole turn-taking process in the conversation excerpt (and not only to a word, phrase, or utterance). This means these games are being played on micro-, meso-, and macrolevels for interactants. Second is developing a particularized solution to the problem above, rather than relying solely on payoffs. The two aforementioned payoffs B and D from the Decoding Game are not absolute; a particularized solution for labeling a message as a microaggression depends on many other unknowable factors influencing the situation. This means to consider what other conversational turn-taking factors might influence what is happening around the word "proof" and the utterance(s) that carry its contents. These issues are addressed beyond the scope of this chapter. For more information about developing a particularized solution for the contents of messages, see reference [17, 19].

Ultimately, the importance of a pragmatic analysis of microaggressive exchanges is to repair discriminatory comments without defaulting to prescriptive, psychotherapeutic responses (e.g., "I" statements) that potential escalate hostility in social situations [7, 10]. Current repair work focuses on a context-free, atheoretical approach that does not attend to how communicators weigh the relational and social contents of their talk and turn-taking strategies [10]. Game theory demonstrates a few ways to formalize a theoretical approach for explaining what people are doing with their communication, how people understand the content of their communication, why people might interpret and internalize messages as microaggressive given particularized contexts, and how to repair microaggressions during a social interaction without defaulting to communicative scripts that do nothing for de-escalating the heightened emotions in a situation [7]. We get a better sense of how and why people respond to microaggressions in the way that they do and how they might avoid repairing such conflicts when game theory is grounding our comprehension of social situations in a generalizable manner.

Conflict of interest

The author declares no conflict of interest.

Author details

Corey Reutlinger
Arizona State University, Tempe, AZ, USA

*Address all correspondence to: creutlin@asu.edu

References

[1] Pierce CM. Offensive mechanisms. In: Barbour FB, editor. The Black Seventies. Boston, MA: Porter Sargent; 1970. pp. 265-282

[2] Torino GC, Rivera DP, Capodilupo CM, Nadal KL, Sue DW, editors. Microaggression Theory: Influence and Implications. Hoboken, NJ: John Wiley & Sons; 2019

[3] Giles H, Reid SA, Harwood J, editors. Language as Social Action: The Dynamics of Intergroup Communication. Vol. 8. New York, NY: Peter Lang; 2010

[4] Sue DW. Microaggressions in Everyday Life: Race, Gender, and Sexual Orientation. Hoboken, NJ: Wiley; 2010

[5] di Gennaro K, Brewer M. Microaggressions as speech acts: Using pragmatics to define and develop a research agenda for microaggressions. Applied Linguistics Review. 2019;**10**(4):725-744. DOI: 10.1515/applirev-2017-0069

[6] Jones T. Refining Microaggression: A Linguistic Perspective [Internet]. 2016. Available from: http://www.languagejones.com/blog-1/2016/9/8/oi6379payz9mb4diadulndc244gq1s [Accessed: September 3, 2022]

[7] Reutlinger C. Towards a game-theoretic analysis for the study of disability microaggressions as a communicative phenomenon [Doctoral dissertation]. Tempe, AZ: Arizona State University; 2021

[8] Wilkes J, Speer SA. Reporting microaggressions: Kinship carers' complaints about identity slights. Journal of Language and Social Psychology. 2021;**40**(3):303-327. DOI: 10.1177/0261927X20966356

[9] Keller RM, Galgay CE. Microaggressive experiences of people with disabilities. In: Sue DW, editor. Microaggressions and Marginality: Manifestation, Dynamics, and Impact. Hoboken, NJ: John Wiley & Sons; 2010. pp. 241-267

[10] Zahn CJ. A reexamination of conversational repair. Communication Monographs. 1984;**51**:56-66

[11] Lilienfeld SO. Microaggressions: Strong claims, inadequate evidence. Perspectives on Psychological Science. 2017;**12**(1):138-169. DOI: 10.1177/1745691616659391

[12] Sue DW. Microaggressions and "evidence": Empirical or experiential reality. Perspectives on Psychological Science. 2017;**12**(1):170-172. DOI: 10.1177/1745691616664437

[13] Habermas J. On the Pragmatics of Communication. Cambridge, MA: MIT Press; 1998

[14] Leyton-Brown K, Shoham Y. Essentials of Game Theory: A Concise, Multidisciplinary Introduction (Synthesis Lectures on Artificial Intelligence and Machine Learning). San Bernardino, CA: Morgan & Claypool; 2008

[15] Beisecker T, Bostrom RN. Game theory in communication research: A reaction and reorientation. Journal of Communication. 1970;**20**(2):107-120. DOI: 10.1111/j.1460-2466.1970.tb00868.x

[16] Gross M. Mathematical Models in Linguistics. Englewood Cliffs, NJ: Prentice Hall; 1972

[17] Parikh P. Language and Equilibrium. Cambridge, MA: MIT Press; 2010

[18] Parikh P. The Use of Language. Stanford, CA: CSLI Publications; 2001

[19] Parikh P. Communication and Content. Berlin: Language Science Press; 2019

[20] Kreps D, Wilson R. Sequential equilibrium. Econometrica. 1982;**50**: 683-894

[21] Switzer JV. Invisible disabilities. In: Disabled Rights: American Disability Policy and the Fight for Equality. Washington, DC: Georgetown University Press; 2003. pp. 168-172

Chapter 6

Education and Games: Teachers' Professional Knowledge in Integrating Digital Games into Instruction in School

Orit Avidov-Ungar and Merav Hayak

Abstract

This chapter addresses the integration of digital games in the education arena. It presents findings from research investigating the perceptions of teachers at different stages of their career regarding digital game integration in teaching and relates to the perceptions of school principals in this context as well. The findings show that teachers at different stages of their careers integrate digital games in different ways and with varying scopes. It was found that school principals hold positive perceptions of integrating digital games into teaching and encourage, leading teachers to disseminate the new idea of incorporating digital games in instruction.

Keywords: digital games, education, teachers' professional knowledge, teachers' perceptions, teachers' professional development

1. Introduction

In recent years, several trends have emerged that have brought digital games to the center stage in the education arena. The first trend is the accelerated development of digital technology that facilitated the introduction of innovative digital technologies and platforms into our world, among them the internet, the mobile telephone, and the tablet [1]. The technologies enabled wide-scale integration and access to digital games on a range of platforms and with relative ease, in the education arena too. A second trend is the increasing popularity of digital games among students for whom they are an integral part of their daily lives [2–4]. Digital games are close to the world of students who generally play them in their free time. Digital games are perceived as having the potential to arouse enthusiasm, interest, and motivation in a school setting as well [5, 6]. A third trend indicates a significant improvement in learning using digital games in contrast to the use of traditional teaching methods [7, 8]. A fourth trend demonstrates the encounter between the students' desire to play digital games and the education system's willingness to offer innovative and experiential learning environments which advance twenty first-century skills while providing solutions that respond to differences between students [3, 4]. The fifth and last trend took place as a result of the Covid-19 pandemic and the mandatory transition of education systems to online learning. This constituted a significant catalyst in accelerating the integration of techno-pedagogical tools, such as digital games, in teaching and learning [9–11]. In its essence, the digital game is an ideal platform for integration into online learning, so as to generate interest and engagement in learning among students.

As a result, the integration of digital games in education systems is gaining momentum in many countries as it carries the promise of innovation alongside the challenge and opportunity to bring about change in the field of education [12–14]. Thus, many education systems in Europe, Asia, the United States, and the Scandinavian countries choose to integrate digital games into teaching and learning [15, 16]. There are differences between the ways those countries have chosen to implement and assimilate digital games into education. This may be due to the absence of an orderly policy, even between one school and another [17]. In Israel, for example, the Ministry of Education promotes wide-ranging initiatives to integrate digital games in teaching, even encouraging competitions between schools. In general, education systems invest a substantial amount of monetary resources in integrating innovations and digital games into teaching and learning processes. This investment is mainly needed for building a technology infrastructure in schools (such as Wi-Fi), payment for games and purchase of licenses (for example, see **Figure 1**), and teacher and school staff training. Nonetheless, despite the many resources invested in integrating innovation and digital games into teaching and learning processes, the difficulties in integrating them leads researchers to argue that these efforts and attempts are destined to fail and that in most cases, innovative technology, including the integration of digital games, will have a minor effect, if any, on instruction, learning, or students' achievements [18–20].

In this chapter we will present the studies we conducted, which address the integration of digital games in schools from the perspective of policymakers, principals, and teachers. We will, in addition, present the practical implications of integrating digital games in teaching for policymakers and proposals for follow-up research in this context. The chapter opens with insights concerning digital games in the education arena and goes on to present research we conducted as well as the research of others in the field.

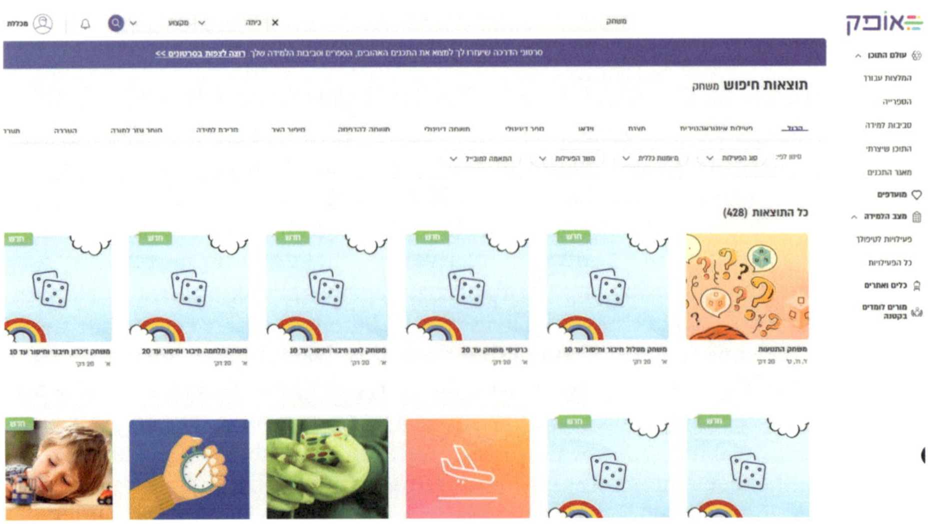

Figure 1.
This screen capture is taken from the "Ofek" website, which offers digital games to Israeli schools for monthly payment (Hebrew).

2. The integration of digital games in the education arena

The research literature includes many definitions of digital games, among them serious games – intended for learning purposes not only in the field of education, video games, online games, etc. Prensky [21] coined the term 'digital game-based learning' (DGBL) and thus distinguished between games whose purpose is entertainment and leisure and games whose aim is teaching and learning. In addition, there are a wide range of games, platforms, and types of games – starting with single-player games to multi-player board games, strategic games, and console games [22]. In this chapter, we will define a digital game as a digital, interactive, and competitive environment that has clear learning goals for promoting teaching and learning processes in education [23, 24].

Many researchers agree that integrating digital games into teaching and learning has everything needed to turn them into effective instructional and learning tools (for example, [5, 25–27]). Digital games enable students to acquire twenty first-century skills including critical thinking skills, research skills, communication skills, self-regulated learning skills, creativity, computing and programming skills, and strategic thinking, which rarely occur in traditional learning [6, 14, 28–31]. These skills are intended to prepare students to optimally adapt to the changes in twenty first-century society and employment.

In contrast, there are researchers who argue that the efficacy of digital games in imparting knowledge and significant learning is controversial and constitutes an obstacle to their integration [22, 32]. Thus, for example, according to Fiorella & Mayer [33], digital games require players to be skillful at identifying and selecting learning tasks, skills that the learner may have not yet acquired. In some cases, this kind of learner autonomy in the game, intended to arouse interest and engagement, may come at the cost of the learning itself [34].

Alongside these findings, certain researchers perceive digital games as a great promise that will lead to a revolution in education [12–14]. However, together with these expectations, there are obstacles to the integration of digital games into education, which reflect the difficulties relating to the teachers' skills, game technology, curricula, costs, etc. [16, 35–41]. The literature, however, rarely relates to these difficulties, using a multidimensional view of the phenomenon from the perspective of the school principals and the teachers who, in practice, lead this change.

3. Main research findings

The research literature concerning digital game integration in teaching and learning highlights the significant role of teachers [15, 16, 29, 42–46]. Several findings point to the importance of focusing on the teachers' perceptions regarding the integration of digital games into teaching and learning. They argue that the negative perceptions of educators, particularly schoolteachers, regarding the efficacy of digital games in teaching and learning may be a barrier to the integration of digital games in practice [17].

Teachers' perceptions regarding the integration of digital games into teaching may derive from the lack of their knowledge concerning the integration of digital games into teaching [47]. Koehler and Mishra [48] proposed a theoretical framework for understanding teachers' Technological Pedagogical Content Knowledge (TPACK). TPACK is a framework for the integration and effective use of technology in the classroom, arguing that teachers need to understand the link between three bodies of knowledge: pedagogy, the content being taught, and technology. The concept is based on Shulman's [49] theory, which argues that teacher pedagogy and teacher content knowledge should not be treated as distinct knowledge areas. Studies that examined the willingness of teachers to integrate digital games into their instruction found that they lacked the necessary knowledge and skills for digital games [15, 50] and, in particular, lacked the game literacy needed to enhance and facilitate digital game integration [51].

In this context, claims that have been made in recent years relate to the absence of the games component from the TPACK framework [52], and importantly, the practice teachers need to integrate digital games. Research we conducted shows that the type of professional knowledge teachers need to integrate digital game-based learning in schools requires diagnosis, intervention, and inference skills. These professional skills are essential in order to respond to each student's and class's particular needs in the promotion of learning achievements. The claim is that teachers have been left on their own to decide how to integrate digital games and evaluate the students' achievements [37].

The results of other research we conducted shows that teachers at different stages of their career integrate digital games differently [37]. Thus, novice teachers in the early stages of their career tend to integrate digital games right as they start out. Teachers at late stages of their career perceive digital games as less effective and integrate them into their lessons to a lesser extent, as compared to other groups of teachers. Teachers at advanced stages of their career perceive digital games as more effective than teachers in the different groups. They perceive themselves as integrating digital games in their teaching to a greater degree than teachers in the different groups. They perceive digital games as an innovative and advanced teaching techno-pedagogical tool that enables them to create inquiry-based learning, differential learning, alternative assessment, and a flipped classroom into their teaching. They prepare for the integration of digital games in a more comprehensive way, as compared to the teachers in the different groups.

Another research study we conducted, which has not yet been published, examined the perceptions of school principals, who lead the integration of digital games in their schools, related to strategies they employ in this context. The findings show that most of the school principals have positive perceptions regarding the integration of digital games into teaching and learning, but they emphasize that teachers play an important role in assessing the contribution of digital games to the students' learning and achievements. In addition, these school principals deploy several strategies to promote the integration of digital games in their school, for example, using leading teachers in the school, the ones who integrate digital games in their teaching, as "change engines" to lead the move to digital game integration. This means that they leverage the leading teachers, who are devoted enthusiasts of using digital games in their teaching, in a number of strategic ways: disseminating success stories, participating in the school's decision-making processes, establishing peer instruction, and finally, defining them as the school's "agents of change" [35, 53]. Thus, the principals believe that by observing leading teachers, other teachers can be positively influenced and can revise their beliefs regarding the integration of digital games [54]. Eyal and Yosef-Hassidim [55] point out in their research on teachers who champion new technologies that they carry the weight of leading technological innovation in their schools and indicate that these teachers collapse under its burden and do not succeed in meaningful diffusion of the technology in their schools. Integrating digital tools, including digital games, is conducted in small "islands," in school [56]. Therefore, the innovation fails in expanding to the point of "total innovation" integrated within all levels of the organization. Nevertheless, the "islands of innovation" do serve schools by allowing them to present to their environment the appearance of having adopted innovation, thus leading them to gain legitimacy along with the resources it brings [57, 58].

Furthermore, the results of our research that examined the factors inhibiting and supporting teachers in integrating digital games into their teaching and learning found that the technical issue is the major factor inhibiting teachers from integrating digital games in their instruction. Teachers noted the technical, logistical, and operational difficulties involved in operating the equipment and the technical infrastructure for using digital games in their classroom as the main inhibiting factor. In contrast, the main supporting factor stems from the influence of enthusiastic teachers, which results from their experience of seeing students' learning experience and engagement in learning using digital games.

Related to the supporting and inhibiting factors for integrating digital games of teachers at different career stages [37], we found differences between the teacher groups. Teachers at early stages of their career mentioned the need to enforce discipline in their classroom as an inhibiting factor. Teachers at advanced and late stages of their career mentioned technical factor as an inhibiting factor. With respect to supporting factors, among teachers in early stages of their career the pedagogical and professional factor was found to be supportive, whereas teachers in advanced and late career stages stressed the effect of the students' positive learning experience using digital games as the main supporting factor. Similar findings were seen in the research literature [35, 37, 59]. These findings indicate the need for teachers' professional development in the integration of digital games in teaching and learning.

4. Practical implications

Based on the research findings and the insights emerging from the studies we conducted, we wish to clarify a number of recommendations regarding two aspects: (1) policy related to the integration of digital games in schools and (2) teachers' professional development.

4.1 Policy concerning the integration of digital games

Research shows that schools are expected to integrate innovative technology, as emerges, for example, from the documents of the European Union Commission, which finances several research projects to promote this issue [60, 61]. In addition, there are programs and projects to integrate digital games into classroom instruction as part of the program to integrate technology [16, 39, 62, 63]. For example, Scotland, Australia, and Singapore operate programs to integrate digital games into the instruction of science and mathematics and programs for developing and designing digital games by students. Recently, in Eastern Asia, emphasis was placed on the integration of innovation in and the use of technology including digital games to teach English [64]. In the United States, the Department of Education declared that it is aware of the effectiveness of digital game use for learning and is therefore committed to widespread and comprehensive integration of high-quality games in schools and in non-formal education settings [65]. In 2013, DigComp, the European

Digital Competence Framework, published a list of digital skills and competencies students need to acquire in the digital age, including digital competency and knowledge needed for using digital technologies, with digital games being counted among the tools used for acquiring these skills [66]. Therefore, we recommend that countries, districts, and schools formulate a clear policy concerning the integration of digital games into the instruction in school. This will align policy with expectations of innovation in schools and define achievement goals for digital competency among the relevant stakeholders.

In our experience, integration of digital gaming is also related to policy. Without policy, implementation is carried out only by "devoted enthusiasts." Our practical experience shows that when leaders of education systems draw up policies regarding the integration of technology, in general, and digital games, in particular, it increases the likelihood of assimilating digital games as part of teaching and learning processes in education systems.

When educators succeed in integrating digital games into their teaching and experience success in this context, there is a high probability that they will continue to integrate them in the future as well.

4.2 Teachers' professional development

Research indicates the need for in-depth understanding of how to support teachers in their role as change agents in a way that will help implement ongoing and sustainable innovation in schools [67]. To this end, professional development related to the integration of digital games is needed throughout the stages of the teachers' careers. Such professional development programs are still in their infancy [68]. Thus, for example, in Europe there are professional development courses for teachers on diagnostic tools that help them to assess the suitability of games to students and to teaching. For instance, since 2014, there are MOOCs (massive open online course) for teachers across Europe that guide them on how to integrate digital games into their teaching. Helping teachers be able to diagnose the degree of a game's suitability to teaching and learning, they use the PEGI (Pan-European Game Information) system for games (see **Figure 2**), in general, and for digital games as well. This system classifies the content of the game and its suitability for the target audience although the rating received is enforced only in Finland and Norway [69]. In Israel, the situation is somewhat similar to what is taking place around the world.

In addition, as mentioned, the findings of the research we conducted among teachers at different stages of their career show that there is a need for in-service education throughout the stages of the teachers' career and aligned with the stage of the teacher's career. For example, in accordance with the teacher's career stage, customized development programs can be offered to interested teachers [70]. In training, it is very important to include practical experience in integrating digital games and development of capabilities to assess learning using digital games [47]. Similarly, training should be accompanied by an evaluation of the teacher with respect to her perceptions and knowledge of technology, in general, and the integration of digital games, in particular. Such evaluation will provide a snapshot of the teacher's knowledge and perceptions. Professional development should begin already at the teacher-training stage, and therefore, having policymakers is important at teacher education colleges/universities to lead this policy among the teaching faculty at the college/university [53, 71]. We recommend devoting thought in advance to the professional development processes needed and as part of the process of integration

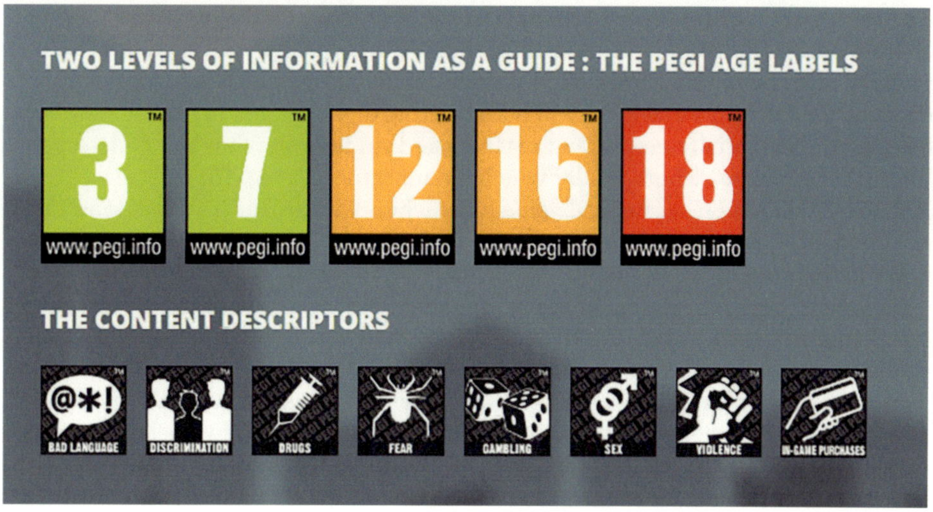

Figure 2.
PEGI (Pan-European Game Information) provides age classifications for video games in 38 European countries. The age rating confirms that the game is appropriate for players of a certain age.

of digital games into teaching and learning, to carefully plan customized development processes for teachers that are aligned with their knowledge, perceptions, and their career stage. All this should be conducted in coordination with their school principals and in line with the policy defined.

In the context of the professional development of teachers, in our experience, professional development is critical for leading change processes in education systems. We know that teachers experience many changes. These should be accompanied by the definition of goals as part of their professional development. A great number of teachers do not have the skills and abilities needed to assimilate technology in teaching or to integrate digital games into their teaching. To succeed in motivating teachers to integrate digital games into their teaching, systematic and orderly development processes for learning the subject must be built. Learning, however, is insufficient; it should be accompanied by support and guidance in the field during assimilation into classroom teaching.

Teachers at different career stages have varying degrees of technological knowledge and require training and guidance tailored to their level - this is what we have learned from our experience in the education field.

4.3 Suggestions for follow-up research

In order to expand the understanding of the integration of digital games in education, it is suggested that research concerning stakeholders in and out of the school setting be conducted, as this may shed the light on teachers' perceptions, adoption and integration, and professional development concerning the integration of digital games in teaching and learning. These stakeholders are school principals, mid-level school management, parents, students, parties in the administration responsible for digital game integration in schools, etc. Regarding teachers' professional development in the integration of digital games, there is a need to deepen the understanding of the teachers' pedagogy when using games in the classroom, meaning that there is a need for a more faithful description of the teachers' activities during the process of the integration of digital games. Consequently, future research can shed light on how teachers integrate digital games into the curriculum and on which teaching-supported practices they should implement when integrating digital games in their classroom. If we learn substantially more about teachers and students in schools and classrooms, about what works for whom, under which circumstances, and with what effects, the knowledge will have real practical value for teachers who are meant to respond to development processes with respect to the effectiveness of the integration of digital games in their teaching in schools [72].

In concluding this chapter, we would like to reiterate that digital games are an important techno-pedagogical tool and investment must be made in its assimilation among teachers as part of teaching and learning processes.

Studies show, as mentioned, that digital games contribute to students' motivation and help to impart twenty first-century skills. As those involved in teacher training and professional development of teachers, we recommend adopting techno-pedagogical approaches in teaching among teachers that will use digital games as part of their teaching.

If, during teacher training, students were to experience using digital games, there is a greater likelihood that they will later use digital games as part of their teaching.

As mentioned, although the use of digital games requires professional development processes and guidance, it leads to diverse, challenging, and enriched teaching for the students.

Conflict of interest

The authors declare no conflict of interest.

Author details

Orit Avidov-Ungar[1*] and Merav Hayak[2]

1 Achva Academic College, Israel

2 Ben-Gurion University of the Negev, Israel

*Address all correspondence to: oritav65@gmail.com

References

[1] Brom C, Šisler V, Slavík R. Implementing digital game-based learning in schools: Augmented learning environment of 'Europe 2045'. Multimedia Systems. 2010;**16**:23-41

[2] Beck JC, Wade M. The Kids are Alright: How the Gamer Generation is Changing the Workplace. Boston, Massachusetts: Harvard Business Press; 2006

[3] Huizenga J, Admiraal W, Akkerman S, Dam GT. Mobile game-based learning in secondary education: Engagement, motivation and learning in a mobile city game. Journal of Computer Assisted Learning. 2009;**25**:332-344

[4] Van Eck R. Digital game-based learning: It's not just the digital natives who are restless. EDUCAUSE Review. 2006;**41**:16

[5] Ke F. Designing and integrating purposeful learning in game play: A systematic review. Educational Technology Research and Development. 2016;**64**:219-244

[6] Kirriemuir J, McFarlane A. Literature Review in Games and Learning. A Graduate School of Education. UK: University of Bristol: Futurelab; 2004. [Internet]. Available from: http://www.futurelab.org.uk

[7] De Freitas S. Are games effective learning tools? A review of educational games. Journal of Educational Technology & Society. 2018;**21**:74-84

[8] Zhonggen Y. A meta-analysis of use of serious games in education over a decade. International Journal of Computer Games Technology. 2019;**2019**:1-8. Article ID 4797032. DOI: 10.1155/2019/4797032

[9] Dubé AK, Dubé NJ. Policies to guide the adoption of educational games into classrooms. Educational Technology Research and Development. 2021;**69**(1):167-171

[10] Reimers F, Schleicher A, Saavedra J, Tuominen S. Supporting the continuation of teaching and learning during the COVID-19 Pandemic. OECD. 2020;**1**:1-38

[11] UNESCO. COVID-19 educational disruption and response [Internet]. 2020. Available from: https://en.unesco.org/themes/education-emergencies/coronavirus-school-closures

[12] Gee JP. What Video Games Have to Teach Us About Learning and Literacy. New York: Palgrave Macmillan; 2003

[13] Kirkley SE, Kirkley JR. Creating next generation blended learning environments using mixed reality, video games, and simulation. TechTrends. 2004;**49**:42-89

[14] Prensky M. Digital Game-based Learning. New York: McGraw-Hill; 2001

[15] Hsu CY, Tsai MJ, Chang YH, Liang JC. Surveying in-service teachers' beliefs about game-based learning and perceptions of technological pedagogical and content knowledge of games. Journal of Educational Technology & Society. 2017;**20**:134-143

[16] Joyce A, Gerhard P, Debry M. How are Digital Games Used in Schools: Complete Results of the Study. Brussels, Belgium: European Schoolnet EUN Partnership AISBL; 2009

[17] Huizenga JC, Ten Dam GTM, Voogt JM, Admiraal WF. Teacher perceptions of the value of game-based

learning in secondary education. Computers & Education. 2017;**110**:105-115

[18] Fullan M. The New Meaning of Educational Change. 5th ed. New York, NY: Teachers College Press; 2015

[19] Allen LK, Crossley SA, Snow EL, McNamara DS. L2 writing practice: Game enjoyment as a key to engagement. Language Learning & Technology. 2014;**18**:124-150

[20] Fuhrman S, Elmore R. Redesigning Accountability Systems for Education. Vol. 38. NY: Teachers College Press; 2004

[21] Prensky M. Digital game-based learning. Computers in Entertainment (CIE). 2003;**1**:21-21

[22] Young MF, Slota S, Cutter AB, Jalette G, Mullin G, Lai B, et al. Our princess is in another castle: A review of trends in serious gaming for education. Review of Educational Research. 2012;**82**:61-89

[23] Erhel S, Jamet E. Digital game-based learning: Impact of instructions and feedback on motivation and learning effectiveness. Computers & Education. 2013;**67**:156-167

[24] Prensky M. Don't Bother Me, Mom, I'm Learning! How Computer and Video Games are Preparing Your Kids for 21st Century Success and how You Can Help! New York: Paragon House; 2006

[25] Chen CH, Shih CC, Law V. The effects of competition in digital game-based learning (DGBL): A meta-analysis. Educational Technology Research and Development. 2020;**68**:1855-1873

[26] Jabbar AI, Felicia P. Gameplay engagement and learning in game-based learning: A systematic review. Review of Educational Research. 2015;**85**:740-779

[27] Wouters P, Van Oostendorp H. A meta-analytic review of the role of instructional support in game-based learning. Computers & Education. 2013;**60**:412-425

[28] Groff., Howell, C, Cranmer S. The Impact of Console Games in the Classroom: Evidence from Schools in Scotland. UK: Futurelab; 2010

[29] Klopfer E, Osterweil S, Salen K. Moving Learning Games Forward. Cambridge, MA: The Education Arcade; 2009

[30] McFarlane A, Sparrowhawk A, Heald Y. Report on the Educational Use of Games. Cambridge. [Internet: TEEM; 2002 Available from: http://consilr.info.uaic.ro/uploads_lt4el/resources/pdfengReport%20on%20the%20educational%20use%20of%20games.pdf

[31] Shaffer DW, Squire KR, Halverson R, Gee JP. Video games and the future of learning. Phi Delta Kappan. 2005;**87**:105-111

[32] Foster AN, Shah M, Duvall M. Game network analysis: For teaching with games. In: Foster AN, Shah M, Duvall M, editors. Teacher Education: Concepts, Methodologies, Tools, and Applications. USA: IGI Global; 2016. pp. 371-403

[33] Fiorella L, Mayer RE. Paper-based aids for learning with a computer-based game. Journal of Educational Psychology. 2012;**104**:1074

[34] Easterday M, Aleven V, Scheines R, Carver S. Using tutors to improve educational games. Proceedings of the 15th International Conference on Artificial Intelligence in Education; Berlin Heidelberg: Springer Verlag. 2011. pp. 63-71

[35] Avidov-Ungar O, Hayak M. Teacher perception of the adoption

and implementation of DGBL in their classroom teaching: Adoption and implementation of DGBL among teachers. International Journal of Game-Based Learning (IJGBL). 2021;**11**:17-30

[36] Baek YK. What hinders teachers in using computer and video games in the classroom? Exploring factors inhibiting the uptake of computer and video games. CyberPsychology and Behavior. 2008;**11**:665-671

[37] Hayak M, Avidov-Ungar O. The integration of digital game-based learning into the instruction: Teachers' perceptions at different career stages. TechTrends. 2020;**64**(6):887-898

[38] Groff J, Clarke-Midura J, Owen VE, Rosenheck L, Beall M. Better learning in games: A balanced design lens for a new generation of learning games. Cambridge, MA: MIT Education Arcade and Learning Games Network; 2015

[39] Koh E, Yeo G, Wadhwa B, Lim J. Teacher perceptions of games in Singapore schools. Simulation Gaming. 2012;**43**:51-66

[40] Papadakis S. The use of computer games in classroom environment. International Journal of Teaching and Case Studies. 2018;**9**:1-25

[41] Piller Y, Roberts-Woychesin J. Raising parental awareness about game-based learning. In: Proceedings of the 7th International Conference on Computer Supported Education-Volume 2-CSEDU (2); SCITEPRESS-Science and Technology Publications. 2015. pp. 385-389

[42] Becker K. Digital game-based learning once removed: Teaching teachers. British Journal of Educational Technology. 2007;**38**:478-488

[43] Bourgonjon J, Valcke M, Soetaert R, Schellens T. Students' perceptions about the use of video games in the classroom. Computers & Education. 2010;**54**:1145-1156

[44] Chee YS, Mehrotra S, Ong JC. Professional development for scaling pedagogical innovation in the context of game-based learning: Teacher identity as cornerstone in "shifting" practice. Asia-Pacific Journal of Teacher Education. 2015;**43**:423-437

[45] Connolly TM, Boyle EA, MacArthur E, Hainey T, Boyle JM. A systematic literature review of empirical evidence on computer games and serious games. Computers & Education. 2012;**59**:661-686

[46] Williamson B. Computer games, schools, and young people: A report for educators on using games for learning [Internet]. Futurelab. 2009. Available from: http://www.futurelab.org.uk/resources/documents/project_reports/becta/Games_and_Learning_educators report.pdf

[47] Allsop Y, Jessel J. Teachers' experience and reflections on game-based learning in the primary classroom: Views from England and Italy. International Journal of Game-Based Learning (IJGBL). 2015;5:1-17

[48] Koehler M, Mishra P. What is technological pedagogical content knowledge (TPACK)? Contemporary Issues in Technology and Teacher Education. 2009;**9**:60-70

[49] Shulman LS. Those who understand: Knowledge growth in teaching. Educational Researcher. 1986;**1986**(15):4-14

[50] Hsu TY, Chiou GF. Preservice science teachers' prior gameplay experience and their perceptions of digital game-supported learning. In: EdMedia+

Innovate Learning, Lisbon, Portugal. Waynesville, NC: Association for the Advancement of Computing in Education (AACE); Jun 2011. pp. 3331-3339

[51] Takeuchi LM, Vaala S. Level up learning: A national survey on teaching with digital games. In: Joan Ganz Cooney Center at Sesame Workshop. New York, NY: Center at Sesame Workshop; 2014

[52] Hsu CY, Liang JC, Chai CS, Tsai CC. Exploring preschool teachers' technological pedagogical content knowledge of educational games. Journal of Educational Computing Research. 2013;**49**:461-479

[53] Hayak M, Avidov-Ungar O. Perceptions of senior academic staff in colleges of education regarding integration of technology in online learning. Journal of Information Technology Education: Innovations in Practice. 2022b;**21**:77-94

[54] Winter E, Costello A, O'Brien M, Hickey G. Teachers' use of technology and the impact of Covid-19. Irish Educational Studies. 2021;**40**:235-246

[55] Eyal O, Yosef-Hassidim D. Managing educational champions: Entrepreneurship in schools. Journal of School Leadership. 2012;**22**:210-255

[56] Avidov-Ungar O, Eshet-Alkalay Y. Islands of innovation: A critical analysis of a model for innovation implementation in school systems. Adult Education in Israel. 2014;**13**:60-90

[57] Avidov-Ungar O. "Islands of innovation" or "comprehensive innovation." Assimilating educational technology in teaching, learning, and management: A case study of school networks in Israel. Interdisciplinary Journal of E-Learning and Learning Objects. 2010;**6**:259-280

[58] Christensen CM, Aaron S, Clark W. Disruption in education. Educause Review. 2002;**38**:44

[59] Hayak M, Avidov-Ungar O. Knowledge and planning among teachers integrating digital game-based learning into elementary school classrooms. Technology, Pedagogy and Education. 2022a (Forthcoming)

[60] Blamire R. Games based learning digital games for learning: Conclusions and recommendations from the IMAGINE project. EuropeanSchoolnet [Internet]. 2010. Available from: http://recursostic.educacion.es/blogs/europa/media/blogs/europa/informes/IMAGINE%20Conclusions%20and%20recommendations%202010-3.pdf

[61] Cagiltay K, Reinhardt R. Green Paper European Network for Growing Activity in Game- based learning in Education – ENGAGE [Internet]. 2010. 2010. Available from: http://www.engagelearning.eu/community/wp-content/uploads/2011/02/WP5_D15_METU_EFQUEL_Green-Paper-on-GBL.pdf

[62] Ellis H, Heppell S, Kirriemuir J, Krotoski A, McFarlane A. Unlimited learning: Computer and Video Games in the Learning Landscape. London: Entertainment and Leisure Software Publishers Association; 2006. [Internet]. DOI: 10.13140/RG.2.2.14632.37127

[63] European Schoolnet. Games in schools [Internet]. 2009. Available from: http://games.eun.org/upload/gis-full_report_en.pdf. Synthesis report available at http://games.eun.org/uploade/gis-synthesis_report_en.pdf.

[64] Butler YG. The use of computer games as foreign language learning tasks for digital natives. System. 2015;**54**:91-102

[65] Office of Educational Technology. Games for learning 2014 (September). [Internet]. Available from: https://tech.ed.gov/games/

[66] Ferrari A. DIGCOMP: A Framework for Developing and Understanding Digital Competence in Europe. Luxembourg: European Union; 2013

[67] Holdsworth S, Maynes N. "But what if I fail?" A meta-synthetic study of the conditions supporting teacher innovation. Canadian Journal of Education/Revue Canadienne de l'Éducation. 2017;**40**:665-703

[68] Nousiainen T, Kangas M, Rikala J, Vesisenaho M. Teacher competencies in game-based pedagogy. Teaching and Teacher Education. 2018;**74**:85-97

[69] Tsekleves E, Cosmas J, Aggoun A. Benefits, barriers and guideline recommendations for the implementation of serious games in education for stakeholders and policymakers. British Journal of Educational Technology. 2016;**47**:164-183

[70] Avidov-Ungar O. A model of professional development: teachers' perceptions of their professional development. Teachers and Teaching. 2016;**22**:653-669

[71] Avidov-Ungar O, Hayak M. Attitudes toward the integration of digital games into instruction in teacher education colleges during the COVID-19 pandemic. Journal of Information Technology Education: Research. 2022;**21**:623-639. DOI: 10.28945/5037

[72] Hargreaves DH. The production, mediation and use of professional knowledge among teachers and doctors: A comparative analysis. In: Knowledge Management in the Learning Society. OECD; 2000. pp. 219-238. Available from: https://books.google.co.il/books?hl=iw&lr=&id=1-uaAwAAQBAJ&oi=fnd&pg=PA219&dq=Hargreaves+DH.+The+production,+mediation+and+use+of+professional+knowledge+among+teachers+and+doctors:+A+comparative+analysis.+Knowledge+Management+in+the+Learning+Society.+OECD.+2000:219-238&ots=kxXDb7dd65&sig=M_leT5sToeVqPsdZEDAV7c2QkNk&redir_esc=y#v=onepage&q&f=false

Chapter 7

Game Development and Testing in Education

Emília Pietriková and Branislav Sobota

Abstract

The following chapter presents an experience with academic education in computer games design and development based on game theory, emphasizing iterative development, rapid (paper) prototyping, and game testing. We demonstrate and comment on various educational activities focused on game testing, describing them in steps and including benefits and motivation. Iteration of card game rules, design of a board game, design of a computer game with a paper prototype, and testing of early-access games are supplemented with students' practical outputs in both text and graphical form. The results include feedback in the testing forms: Focus Groups, Playtesting, Usability Testing, and Quality Assurance.

Keywords: game design, game theory, game testing, playtesting, education

1. Introduction

In addition to forming a game development basis, Game Theory can significantly affect the development of computer games. Some types of games have been known in Game Theory for a long time. However, some procedures are modified concerning modern technologies, allowing one to quickly solve some (up to now) complex tasks, e.g., independently of the computer graphics use. It is excellent when some principles from this theory are applied in the education process of computer games development, especially in the area of computer science education.

In terms of influence (**Figure 1**), we can observe mainly the following approaches to computer games testing:

- *Direction GTh (Game Theory) → GTe (Game Testing)*: The Game Theory formalism defines the framework of Game Testing. It can determine what to

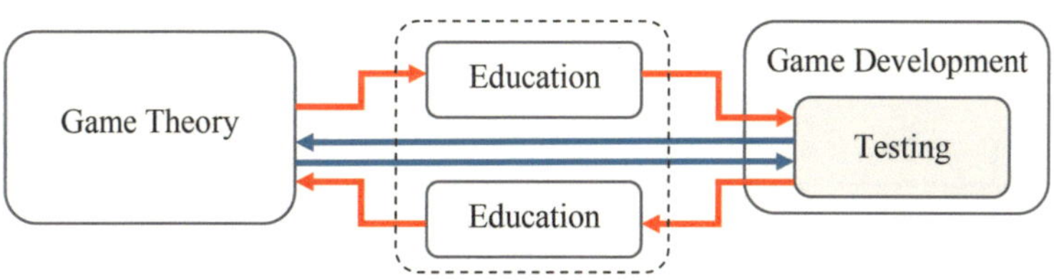

Figure 1.
The effect of game theory on game development and vice versa.

focus on during testing, notably with a specific type of game or what testing parameter to use. The formalism can specify possible threats and other aspects when including a SWOT analysis.

- *Direction GTe (Game Testing) → GTh (Game Theory)*: In this case, the testers define the game evaluation parameters by recording their notes (text, voice, video), which are subsequently processed and structured. Game designers and developers can afterward perform simulation and evaluation to optimize the game, game loop (**Figure 2**), improve its design, or change the formal apparatus. In addition, it is possible to obtain feedback within the game loop for inputs or outputs adjusting, including the possibility of adjusting the implementation of artificial intelligence, e.g., in the form of boots. The game loop implemented in this way also makes it possible to apply separate/simultaneous testing of individual blocks, and thus it increases the efficiency and speed of the testing itself. It also affects better processing, e.g., incidence matrix. In a game loop structured in this way, e.g., at the input level, it is also possible to implement various scripting languages, in the case of inserting automatic test scripts during game development. Of course, the elements of the Game theory are mostly applied in the System kernel. This process can help compare to what extent the game design corresponds to the formalism of Game Theory (Is the game playable, is it difficult, is the game and its reward strategy motivating, or does the symmetry of the game change?).

- *Direction Ed (Education) → GTe and GTh:* Determines how to teach *CS* students, so they can appropriately use the *GTh* knowledge in *GTe*, and subsequently in the practical development of computer games. This direction substantially impacts students in the sense of the *GTh* education and gamification, so they are not only "flooded" by the mathematical apparatus but can also transform it into programming practice.

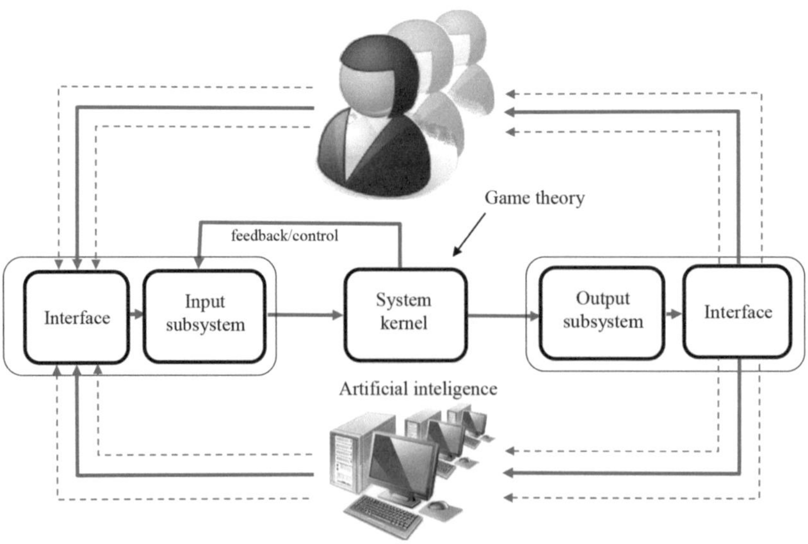

Figure 2.
Standard process of player(s) interaction in game loop.

Regarding teaching computer games development, some *GTh* results can help define roles from a *GTe* perspective (e.g. Observer, Playtester, Bugfixer). Each one may have its formal apparatus or different parameters (values, methods of use, or semantics), depending on how the result supported by the theory applies to individual GTe roles, rules, strategies, or incidences.

At the Department of Computers and Informatics FEEI Technical University of Košice, we have implemented the *Design and Development of Computer Games* and related courses for several years. To be successful in implementation, the *CS* students, as future creators of computer games, need to have at least the basic apparatus of Game Theory. As part of the education process, the students become familiar with various methodologies and technologies related to computer games design and development: high concept, market research, storytelling, prototyping, rules balance, MDA, or iterative game development, an essential part of which is *GTe*.

In the chapter, we would like to present the primary forms of this education process, focusing on *GTe*, including some of the technologies used. We will also present some forms and demonstrations of *GTe* performed by students, corresponding mainly to the first two directions. We processed some results with an SUS questionnaire using a 5-point Likert scale.

We will present activity outputs recorded during the actual education process with 3rd-grade students of the bachelor's degree in Informatics in a full-time attendance form of study. We believe the experience presented in this chapter will find its readers. We attach great importance to the fact that these are actual data processed over multiple years and thus can provide a basis for other studies. We will comment on the outputs of student activities from the educational process point of view.

1.1 Iterative game development and testing

Testing is one of the critical processes accompanying the design and development of computer games. Game designers and developers often confuse or simplify different testing processes, leading to unsatisfactory results, e.g. for students who encounter testing as part of the educational process, it is most often confusion of *playtesting* with *bugfixing* (*quality assurance*) or playtesting with *usability testing* (*interface*). Therefore, in the next chapter, we will formally clarify the testing process and present one of the ways to incorporate various forms of testing into the educational process in computer game design and development.

Well-known literature states that four types of testing are crucial to the design and development of computer games [1, 2]:

- *Focus Groups:* They play a role, especially in the initial stage of game design, consisting of meetings with potential players providing feedback, likes, and dislikes on the upcoming game topic. The literature [1] states that although Focus Groups can be helpful, they are unpopular. The reason is management's fear of "killing" their ideas, so the book [2] recommends switching from obtaining one-sided feedback to discussions generating new ideas. Such Focus Groups will allow iterating on existing ideas more effectively. Moreover, it does not have to be only the initial stages of game design. They can also be helpful later, e.g., when designing game levels or planning the marketing of an upcoming game, which is especially crucial in the final stage of game development.

- *Quality Assurance (QA) Testing:* This is standard bugfixing. The goal is not to play and enjoy the game but to identify all the problems and make sure the game meets all the predetermined requirements [3]. QA engineers look for potential bugs and report them. Since games are not a common type of software, a video, image, and detailed description can help understand steps, leading to the error situation.

- *Usability Testing:* One of the final aspects of a tested game is accessibility. Usability Testing focuses on whether the user interface is intuitive and easy to use. Players must understand and play the game without assistance. For this reason, testers best test the game during the first-time access [2]. Ideally, User Experience (UX) experts perform the testing.

- *Playtesting:* Often referred to as the most important of the four testing types. It is a process that designers and playtesters carry out throughout the game design to determine how the players experience the game. The team of designers and developers proves they are making a good game, even if it scares them [1]. Playtesting aims to guide the design and provide feedback to the designers on whether the game meets their goals and the players' expectations [4]. The forms are different, more or less formal.

Game User Research (*GUR*) intends to improve player experience [4]. *GUR* is a community of UX experts, game developers, and researchers, where the primary research subject is playtesting. In general, *GUR* supports the adaptation of standard HCI evaluation techniques in the playtesting practice, focusing on "entertainment" instead of "productivity" [5]. From the point of view of software engineering methods, *GUR* mainly recommends the *RITE* method [6]. The research applies other sophisticated methods such as *PLAY*, adapting various heuristic methods [7], or using artificial intelligence - AI Players. The goal is to predict player behavior and gaming experience [8].

As already mentioned, testing is critical in the iterative process of game design and development. It helps detect problems early (it is still possible to fix them). But what exactly is an iterative process?

Rapid Iterative Testing and Evaluation (RITE) is a software development method whose main philosophy is to identify a problem as soon as possible and eliminate it. During the last two decades, the *RITE* method has become widespread, especially for products changing rapidly e.g. web, cloud products, and games [6, 9].

It is too late to test at the end of game development. Otherwise, the changes would be too lengthy and costly. The *RITE* method guarantees early identification and elimination of the problem [2]. It is mainly adopted by larger game studios and recommended by the *GUR* community.

There are multiple testing types in game development. Playtesting is key to the *RITE* method, although other types of testing also find their place in individual stages of game design and development (**Figure 3**). From a practical point of view, it is best for playtesting if playtesters can play the game as soon as possible. Rapid prototyping (including paper prototypes) is advantageous if the game is still in the design stage.

A prototype is a simplified but testable game version. Designers create it before they spend money on implementing game elements and features. In general, there are three types of game prototypes [10]:

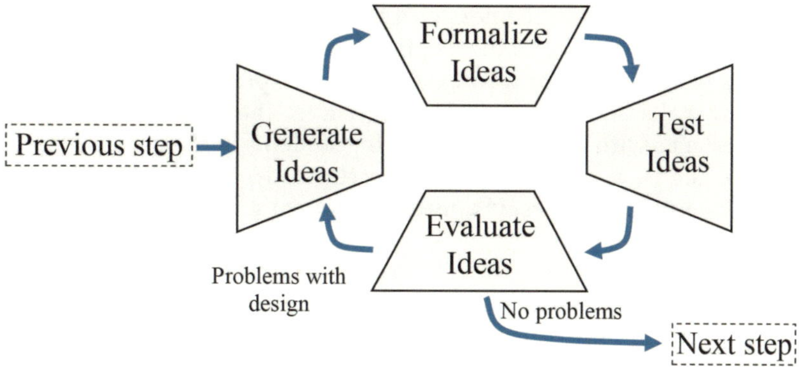

Figure 3.
Process of iterative game design.

- *Software:* Prepared in a game software like Unity,
- *Physical:* Physical interaction and activity with the playtester, and
- *Paper:* Board game, the most common type.

With smaller game industry studios, we encounter risk, so the game is published as soon as possible. In this case, they tend to skip the prototype and playtest only at the end of game development. Many rely on the fact that most game elements are copies of existing elements from other games. This risk results in the publication of many small, imperfect games. Fortunately, the game industry is changing thanks to the GUR community.

For playtesting, which is one of the game testing types, two people are essential:

- *Investigator (Observer):* Person who manages the test,
- *Playtester:* Person who participates in testing and provides feedback.

The Investigator should talk to the players as little as possible to not influence their evaluation and create a questionnaire. The *Investigator* can track the answers' development over time after further iterations. A good *Investigator* should be able to identify playtesters' exit points and not be sad if they do not like the game ideas. The author of [1] recommends a scale of 1–5 instead of 1–10 or 1–7: *Terrible, Pretty bad, So-so, Good,* and *Excellent.*

On the contrary, the *Playtesters* should think out loud while playing, so the *Investigator* can record their thoughts. They should reveal their prejudices and self-analysis. So, in addition to evaluating something positively or negatively, they should also state the reasons. Good *Playtesters* should be able to comment on various game elements like graphics, music, story, mechanics, etc.

2. Game testing in education

As we have already mentioned, in the education process during game testing, we often encounter students confusing playtesting with *bugfixing* or *playtesting* with *usability testing*. Therefore, in the next chapter, we will introduce multiple educational activities in designing and developing computer games, focused on implementing various forms of testing. The goal is to provide students with an experience and make them aware of the differences. The activities proceed from our personal experience with software engineering students in the course of Computer Games Design and Development. We assume the students have already mastered software engineering principles, including traditional software design, development, and testing.

In the following subchapters, we will present multiple outputs of student testing activities. Testers' opinions and feelings are part of testing (especially playtesting). Therefore, the listed results also contain the opinions and feelings of selected students and may not always coincide with our opinions and feelings.

2.1 Card game: playtesting & focus groups

Since computer game design and development correspond to Game Theory, it is appropriate in the education process if students work with typical card or board games. Analysis and playtesting of card games are especially suitable at the beginning of such courses.

Goal: To achieve the best possible gaming experience in the card game.
Resources: Card games (or board games).
Testing: Playtesting & Focus Groups.
Methods: Basic rules iteration, incidence matrix, SWOT analysis.
Length: 80–90 minutes.

Steps:

- Students form teams of 3–5 players.
- Each team selects one existing card game, which they play for about 20–30 minutes. If the game is fast, teams can play it multiple times in a row.
- Group analysis of the game and feedback on the game experience. Here are some steps we recommend:
    - Write a list of card game rules.
    - Write short feedback on the gaming experience, including both positives and negatives.
    - Write SWOT analysis.
- Group iteration of the basic rules to improve the gaming experience.
- Shuffle teams. In each group, 1–2 original players will remain, and 2–3 will shuffle with other groups.
- New teams play card games with new rules.
- Group analysis of the game and feedback on the game experience:

- o Write short feedback on the gaming experience, including both positives and negatives.
- o Update a SWOT analysis.
- o If necessary, prepare a new iteration.

- Compare game experiences both before and after game rules iteration.
- Compare SWOT analysis both before and after game rules iteration.

Feedback on the card gaming experience prepares students for playtesting computer games. Since card games are usually short and fast, students are often unaware of various game aspects, subsequently missing in their feedback. That's why we decided to do a SWOT analysis, the use of which is unconventional in this case. From our experience, the SWOT analysis multiplies the effect of the entire activity.

Since the activity consists of group steps, we can also talk about training Focus Groups. Here we follow the recommendation of [2] that Focus Groups participate in creating new ideas. So, in addition to feedback, students in teams prepare iterations of game rules, coming up with new ideas.

Working with a card game guides students from Game Theory to Game Design. At the end of the activity, they compare playtesting and SWOT analysis both before and after the basic rules' iteration. It points to the quality of the iterations with which the students will develop their evaluation. An alternative could be the same activity with an existing board game.

In the next subchapters, we will present multiple outputs of this activity.

2.1.1 Bang!

Basic rules: Bang! is a wild-west themed card game for 4–7 players in the basic edition. The goal is to kill the enemy according to the roles drawn at the beginning of the game. The sheriff must eliminate all the bandits and the renegade with the help of his assistants. The bandits must kill the sheriff before he kills them. The renegade wants to become the new sheriff, so his goal is to be the last one alive. So, he first takes out all the bandits, sheriff's assistants, and finally, the sheriff. While the sheriff is the only role revealing his identity at the beginning, the other roles hide their identity until they die.

Positive playtesting comment: Can entertain multiple players. It is diverse and strategic.

Negative playtesting comment: The game is difficult for newcomers. It can seem long-winded.

SWOT analysis:

	Useful attributes	Harmful attributes
Internal attributes	• The rules are easy to learn. • If players understand the rules, the game is easy and intuitive. • The mistake should not occur since all players can see the steps of teammates and opponents. • Easily portable game. • It supports strategic and team thinking.	• Needs at least four players. • Some characters from the base pack are not entirely balanced.

	Useful attributes	Harmful attributes
External attributes	• The game with 6–7 players is more entertaining. • Cards quality (material). • If any card is lost, the game is still playable. • Add-on packs make the game more exciting and complement it with existing and new cards. • The possibility of making players' own cards.	• The game with four players becomes repetitive after a while. • Eliminated players can get bored.

Iterations:

1. *Four players:* In the basic game, it is easy to find players' roles after a while. There is the sheriff, vice, and two bandits. The sheriff will show his role before the game starts, so it is easy to determine who are a bandits and who are not. An alternative could be to shuffle the cards by adding a renegade instead of a bandit. The renegade must be the last player left in the game, so he is expected to switch sides

   Playtesting: Iteration made the game harder. It was not immediately clear who had what role.

2. *Three players:* When the sheriff is against two bandits, the player of the sheriff is at a significant disadvantage so that he would have multiple advantages in addition to the original game. He would have two more lives (in the base game, the sheriff always has one more life). Likewise, he would draw one more card (in the base game, it is two cards)

   Playtesting: After the iteration, the sheriff player was no longer at a disadvantage, which positively affected the game balance.

3. *Zombie Bang!* After his death, a person becomes a zombie. He has the same traits and cards but joins the person who killed him. This way, the game will speed up considerably and become more entertaining. Even eliminated players can continue to play

   Playtesting: The game had more dynamics after the iteration. However, there was a problem that the game was too fast.

4. *Gangs:* Players remove the original roles and divide into two or more gangs that will fight against each other. Each has its leader, hidden from others and known only to his fellow members. The goal of the gang is to kill the leaders of the other gangs. If the leader dies, the entire gang automatically loses

   Playtesting: This iteration allowed multiple players to play the game. However, with a smaller number of players, the game became repetitive quickly. Therefore, the minimum recommended number of players for this iteration is 6.

5. *New Sheriff:* If the renegade manages to kill the sheriff before the game ends, he can become the new sheriff, and the game continues. This rule can be extended, for example, that the sheriff's assistants become new renegades, or the original sheriff becomes the new renegade

   Playtesting: Iteration finds its use in case the game ends quickly. After the iteration is applied, the game becomes longer and continues to be dynamic, which was the goal of the iteration.

6. *Deathmatch:* An alternative could be a form of a deathmatch where all players would play against each other regardless of role, which is unnecessary in this case

   Playtesting: The game was more dynamic after the iteration.

7. *Remove weapon restrictions:* The players remove the original condition to fire only at players within a given range

   Playtesting: With this change, we increased the game dynamics and shortened its duration.

SWOT analysis after iterations:

	Useful attributes	Harmful attributes
Internal attributes	• The rules are easy to learn. • If players understand the rules, the game is easy and intuitive. • Easily portable game. • It supports strategic and team thinking.	• Needs at least four players; in some versions, at least six players.
External attributes	• The game with 6–7 players is more entertaining. • Cards quality (material). • If any card is lost, the game is still playable. • Add-on packs make the game more exciting and complement it with existing and new cards. • The possibility of making players' own cards.	• Eliminated players can get bored.

2.1.2 Black Peter

Basic rules: Game for min. 3 players, 33 cards: 32 + 1 (Black Peter). The aim is to collect pairs of identical cards. The players then draw random cards from the left player. The player who has only Black Peter left in his hand loses the game.

Positive playtesting comment: The game is simple for players of any age, has a fast course, and offers a balance of chances.

Negative playtesting comment: When playing, we revealed low variability of the gaming experience along with high repetitiveness and the impossibility of tactics.

SWOT analysis:

	Useful attributes	Harmful attributes
Internal attributes	• A robust rule checking mechanism. • Portable and re-playable. • Simple rules for a different number of players. • A quick game for different ages.	• It is possible to play the game endlessly since the selection of cards by the players is random (for example, when there are two players with three cards, they can move one card around). • The game is repetitive.
External attributes	• It is not necessary to concentrate on the game, and it is possible to have a conversation during the game. • It is possible to play the game even if a card in the pack is lost. • It is possible to play the game with any deck of cards (Pharaoh cards, Joker cards).	• Due to the randomness of the game, one player can lose repeatedly. • Cards can get damaged.

Iterations:

1. *Action pairs:* Upon completing a pair, the player gets the option to "unload" it and perform an action that the pair allows. A player can only unload pairs after the move.

    - Cards with a blue background have essential functions without actions.
    - Cards with a purple background will change the game direction.
    - Cards with a yellow background will allow a player to exchange one of their unwanted cards with any player's card.
    - Cards with a green background will move all cards in the player's hand in the game direction.

    Playtesting: New rules have increased the dynamics of the game. We found a problem where their given action was not always suitable to perform after unloading a pair.

2. The rules changed, so the player does not have to use a particular action after unloading an action pair

    Playtesting: The game became calm. However, players were afraid to use the actions.

3. *Suggested action*: Draw one card from each player. The proposed change includes a new type of action pair with a red background, whose particular action the player must perform. These actions will be deliberately designed to be disadvantageous and annoying to players

    Playtesting: We were pleased with the new dynamics and rule change, and the game provided an entertaining gameplay experience.

SWOT analysis after iterations:

	Useful attributes	**Harmful attributes**
Internal attributes	• A robust rule checking mechanism. • Portable and re-playable. • Simple rules for a different number of players. • A quick game for different ages. • Players need to think more.	• Small children may have a problem with such a game (they may forget some of the rules).
External attributes	• The game is more fun. • It is possible to play the game even if a card in the pack is lost.	• Cards can get damaged.

2.2 Board game: playtesting & focus groups

Board games are suitable as introductory activities in game design courses, like card games. This activity is similar to the card game activity. This time, students are not working with an existing game but creating a new, custom board game that they play, analyze, and iterate with each other.

Goal: To achieve the best gaming experience when designing a board game.

Resources: Paper, pencil, figures, cubes.
Testing: Playtesting & Focus Groups.
Methods: Game design, iteration of basic rules, incidence matrix, SWOT analysis.
Length: 80–90 minutes.

Steps:

- Students form teams of 3–5 players.
- Each team designs and creates a board game on paper. The team can play the game; however, it is not crucial to this activity.
- Each team writes the rules of the board game.
- Shuffle teams. In each group, 1–2 original members will remain, and 2–3 will shuffle with other groups.
- New teams play board games for about 20–30 minutes. If the game is fast, teams can play it multiple times in a row.
- Group analysis of the game and feedback on the game experience. Here are some steps we recommend:
    - Write short feedback on the gaming experience, including both positives and negatives.
    - Write SWOT analysis.
- Group iteration of the basic rules to improve the gaming experience.
- Shuffle teams. In each group, 1–2 original members will remain, and 2–3 will shuffle with other groups.
- New teams play board games with new rules.
- Groups analysis of the game and feedback on the game experience:
    - Write short feedback on the gaming experience, including both positives and negatives.
    - Update a SWOT analysis.
    - If necessary, prepare a new iteration.
- Compare game experiences both before and after game rules iteration.
- Compare SWOT analysis both before and after game rules iteration.

Working with a board game guides students from Game Theory to Games Design, like card game activities. Students practice playtesting a board game design

(**Figure 4**), which is similar to playtesting a computer game design. Again, the SWOT analysis multiplies the effect of the entire activity.

Since the activity consists of group steps, we can also talk about training Focus Groups. At the same time, working with hand-made board games is similar to playtesting with paper prototypes of computer games, preparing students for the following course stages, which are focused on designing and developing computer games.

At the end of the activity, students compare playtesting and SWOT analysis both before and after the basic rules' iteration, developing a self-evaluation again.

2.3 Rapid prototyping: playtesting

This activity moves beyond Game Theory and directly relates to Game Design. After the initial processes like High Concept or Pitch, the game environment, rules, or character design comes next. Since changing these aspects of the game at a later development stage is quite expensive, it is essential to playtest the game as soon as possible.

Goal: To achieve the best possible gaming experience when designing a computer game.

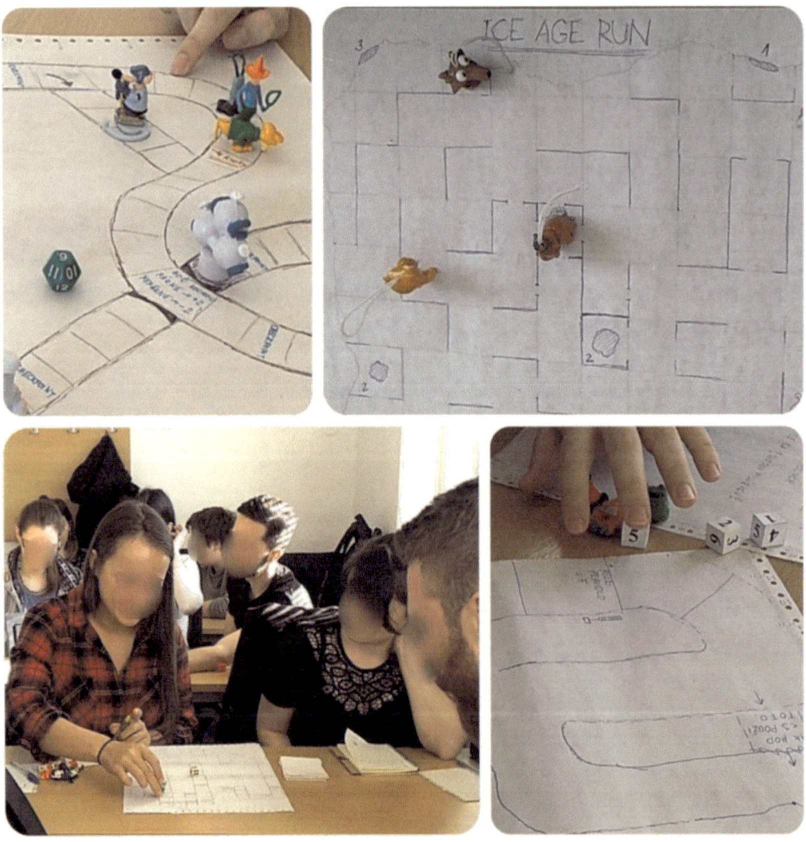

Figure 4.
Board games activity in the classroom.

Resources: Paper, glue, scissors, crayons, figures, Lego.
Testing: Playtesting.
Methods: Game design, basic rules iteration, incidence matrix, paper prototyping.
Length: 180+ minutes.

As mentioned earlier, prototyping is part of computer games' iterative design and development. Students must realize the importance of this stage through the paper prototyping activity.

Steps:

- Students work in existing teams of 3–5 designers and developers.
- Each group already has the basic game proposal, consisting of basic rules, characters, and environment, all prepared during some previous activities.
- Each group prepares a paper prototype for the proposed game (60–120 minutes).
    - Work with paper, glue, scissors, and crayons.
    - Students' creativity is welcome, e.g. they can use various figures or Legos.
- Shuffle teams. In each group, 1–2 original members will remain, and 2–3 will shuffle with other groups.
- New teams play prototype games for about 30 minutes.
- Group analysis of the game and feedback on the game experience. Here are some steps we recommend:
    - New team members are playtesters and they play the game. They talk out loud.
    - The original team members have the role of observers. They talk as little as possible.
    - Write short feedback on the gaming experience, including both positives and negatives.
- In their original composition, the teams analyze the playtesting results.
- In their original composition, the teams iterate aspects of the initial game design.
- If there is enough time, students can update the prototype and playtest the game again.

Testing is an integral part of the software life cycle. In the case of computer games, the design and development process is diverse from traditional software. We emphasize iterations, testing, and evaluation at each stage of development (RITE). A prototype is a practical means of testing, enabling one to carry out the playtesting before the game development begins. Preparing a prototype is faster than implementing an actual computer game, so (not only) the students can get to playtesting much earlier (**Figure 5**).

Feedback on the prototype's gaming experience helps students iterate their computer game's design (**Figure 6**). Reminding individual teams not to be disappointed by negative feedback is significant as one of the playtesting goals is to help game designers but not to demotivate them. Identifying the playtester's "leave" is crucial since we do not want to know that something is good or bad but also why it is good and bad. So, in addition to thinking out loud, the playtester should reveal his prejudices and adequately justify each impression.

In this way, in addition to creating paper prototypes, students practice playtesting computer games, where they have the opportunity to work as both playtesters and observers.

2.4 Early-access games testing: playtesting, QA, UX

This activity aims at testing games at the end of their development. Each student should work as both a tester and an observer, that is, to try testing from both sides. Since three types of testing play a role at the end of game development, students work with all three: *playtesting*, *usability testing*, and *bugfixing* (QA).

Figure 5.
Paper & Lego prototypes activity in the classroom.

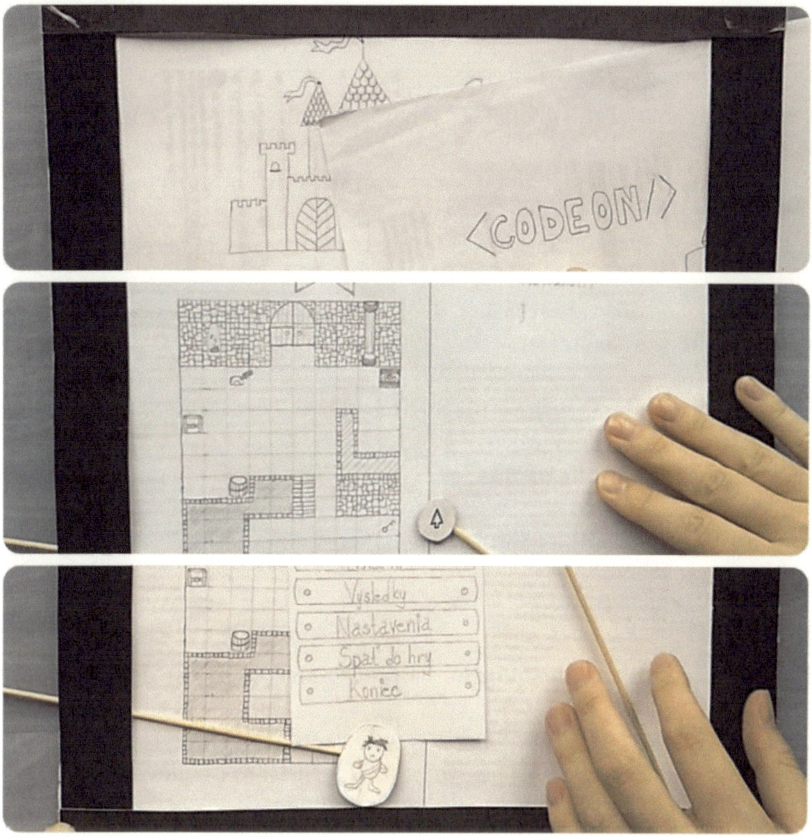

Figure 6.
Example of game screenplay on paper prototype (Slovak menu).

Goal: Practice the role of playtester and observer, achieve the best gaming and user experience, game debugging.
Resources: Computer, computer game.
Testing: Playtesting, Usability Testing, Quality Assurance (QA).
Methods: Questionnaire (SUS), thinking out loud, reporting.
Length: 180–240 minutes.

This activity can have two versions. If students create their games on the course, they test them among each other. If the course has an early-access game available, students test it and report it to the game developers. If possible, students get feedback on their testing from real game developers.

Steps in playtesting of early-access games:

- Students work in pairs.
- One student is an observer; the other is a playtester.
- Playtester prepares an information sheet with helpful information for the game authors:

    - Technical parameters of the OS, computer, and other devices such as resolution, etc.
    - Self-Persona of the playtester (characteristics, self-analysis).
    - Students will also use the information sheet for other types of testing.
- A playtester plays the game and comments on his gaming experience out loud.
- The observer records observations from the gaming experience and does not interfere with the playtester.
- Together, the students write a playtesting report containing:
    - The positives of the gaming experience.
    - The negatives of the gaming experience.
    - Other observations.
- The playtesters fill out the questionnaire if requested by the development team.
- Each playtesting is different, which is why each playtesting is helpful for developers in another way. Therefore, feedback from the developers is essential (if they are willing, which is usually not a problem).

Steps in playtesting of student games:

- Students work in pairs or threes.
- One student is an observer and belongs to the game development team.
- One or two students are playtesters and do not belong to the development team.
- Playtesters play the game and comment on their gaming experience out loud.
- The observer records observations from the gaming experience and does not interfere with the playtesters.
- The development team analyzes the playtesting and prepares new game iteration based on the results.

Steps in usability testing:

- The procedure is similar to playtesting, focusing on the user interface.
- In the case of a student game, the development team prepares a SUS questionnaire with a Likert scale of 1–5.

- Testers add observations related to the game usability to the existing playtesting report.
- Testers prepare a video or an exact sequence of steps to describe problem situations better.
- The testers fill out the questionnaire if requested by the development team.
- In the case of a student game, the development team analyzes the testing and prepares a new game iteration based on the results.

Steps in bugfixing:

- There is no need to work in pairs or speak out loud in this part.
- Each student prepares an information sheet with technical parameters of the OS, computer, and other devices such as resolution, etc.
- Each student independently goes through the game multiple times and tries to interact with as many game elements as possible.
- Each student prepares a report and, if necessary, supplements it with a video or an exact sequence of steps in problem situations.
- In the case of a student game, the development team analyzes the testing and debugs the game implementation.

The essential part of this activity is that students practice the three types of game testing, being on both sides: testers and developers. So, students implement early-access game testing in a broader context.

We have already described the essence of playtesting. The observer should be able to identify the player's "leave" but should not be offended by any negative feedback. On the contrary, the playtester should think out loud, be able to reveal his prejudices, and justify why he likes or dislikes something about the game.

Students can test the game in an early-access version if one is available. Or they can test their games with each other. It all depends on how the educational course is structured.

In the next subchapters, we will present multiple outputs of this activity.

2.4.1 Testing of early-access game bushfires: animal rescue

This chapter presents an example of game testing of the early-access game carried out in 2021. The game is currently available on the Steam platform.

Game: Bushfires: Animal Rescue.
Developer & Publisher: Flying Butter Games.
Version: Early-Access.

Playtester 1, male:

- *Characteristics:* I consider myself 1/3 Achiever and 2/3 Explorer. I'd say I'm a Mid-core gamer, and I like a big open world to explore. I like games where the player has to fulfill various tasks within a level. While playing, I want to logically think about how to pass the level with the highest score, and I enjoy the high-level difficulty.

- *Self-analysis:* For me, the important thing when deciding whether to buy/play a game is whether the game has an idea (it wants to draw the player's attention to a burning problem or sell the player a vital truth). I appreciate it when the developers leave a section in the game where they summarize the information about the theme on which they based the game (for example, in the game Brothers in Arms: Road to Hill 30, the developers left a database of official reports from military missions, according to which they created individual levels/ missions in the game).

- *Top games:* Assassin's Creed: Brotherhood (Ubisoft/Ubisoft Montreal), Euro Truck Simulator 2 (SCS Software), LEGO Pirates of the Caribbean: The Video Game (TT Games Limited), Sims (all series, Electronic Arts), Brothers in Arms: Road to Hill 30 (Gearbox Software/Ubisoft), Blitzkrieg (Nival Interactive/1C Company).

Device	Resolution	OS	Memory	Graphics	Processor
PC	1920 × 1080	Win 10 (x64)	8 GB DDR4 RAM, 256 GB SSD	nVIDIA GeForce GTX 1050 2GB	Intel Core i5 7300HQ 2,5 GHz

Playtester 2, female:

- *Characteristics:* I'm not a regular computer gamer, which means I play a game once a month, sometimes not even that. I cannot get obsessed with games and can play a game for 30 minutes. If the game is multiplayer, I can play it longer without getting bored. Most of the games I would play are from my "childhood," that is, older games. I do not like games where I only have to get from point A to point B.

- *Self-analysis:* If I must choose a new game to play, the visuals and overall graphics of the game and the sound of the game are essential to me - examples are Limbo or Ori And The Blind Forest.

- *Top games:* The Sims 1,2 (Electronic Arts), NHL 06 (EA Sports), Limbo (Playdead), Zoo Tycoon 1 (Blue Fang Games), Worms (Team17).

Device	Resolution	OS	Memory	Graphics	Processor
Note-book	1920 × 1080	Win 10 Home (x64)	8 GB DDR4, 128 GB M.2 SSD	NVIDIA GeForce 930MX/2GB	Intel Core i7 8550U

Playtesting:

- Positive remarks:
    - A fascinating idea for the game. I like that it also points to a situation that happened and shows the players what Australia had to go through.

- It's great that one can also find basic information about animals that have already been discovered and rescued.
- Nice graphics. I like the animals look pretty realistic, and the environment is also very nicely depicted.
- I like the road signs, navigating where to go or what animals are on the road.
- Pleasant music that is not annoying.
- Sound effects when painting a car.

• Negative remarks:

- I find the first map (level) more challenging, considering it is the first contact with the game. It would be nice if the first level was in the form of a tutorial or had an infinite time so the player could master the controls in peace.
- I would not use the sound when selecting the level in the case of the locked ones.

• Other remarks:

- The color design gives a person the impression of dryness and fire.
- The game could include some checkpoints, although they might be difficult to implement now since time is an essential factor, and they could make it too easy to pass the level.

Usability testing:

• At first, the car was sensitive to the controls (reminiscent of the ATV controls in the game GTA: San Andreas), but after one gets used to it, the game has a pleasant control. It is nice that it is possible to use both the keyboard and the mouse.

• I like that when hunting animals, time slows down, so I can better predict where the animal will probably go and think where to shoot the net.

• The colors of the animals are remarkably similar to the environment, which makes them difficult to see.

• It is more difficult to find if the animals are in the fire. It would be good if the animals were marked near the car, just like in the distance, and show the distance from the vehicle. As long as they do not move and the car gets close, the marker indicates where the pet is, and the player knows where to look for it, but once the pet gets closer and moves, it's harder to find (**Figure** 7).

• I'm missing a minimap somewhere in the corner that would speed up navigation. Add a minimap like in GTA or Mafia games, where animal icons would be displayed showing where the animals are located.

Figure 7.
Example of bushfires game - unmarked animals.

Bugfixing (QA):

- *Bug 1 & steps:* When I open the settings during the game and use the ESC key when going back to the game, it does not bring me back to the game, but it brings me back to the settings while I can no longer access the settings via SETTINGS. To go back to the game, I must go back to the menu via BACK from the settings and choose RESUME GAME or return from the settings to the menu and choose RESUME GAME; then it throws me back to the settings, and I choose BACK. (Students attached video).

- *Bug 2 & steps:* In level 4, the target and arrow are not visible in some places while shooting the net to catch the animal. It happens, for example, between two bridges (kangaroos) or at a fence under a hill (koalas). What is interesting is that in these places, the target is not visible if the car is going in the right direction; however, if the vehicle is going in the opposite direction, the target is visible. (Students attached video).

- *Bug 3 & steps:* In level 4, animals disappeared from the map several times. In the video, you can see how the koala disappeared and how the mark of the dingo dog is displayed at first, but then it disappears, and even after leaving, it is impossible to see his mark. (Students attached video).

- *Bug 4 & steps:* I accidentally drove the car through the texture: Over the fence on the bridge (level 3) and through the bridge (level 4). (Students attached video).

- *Bug 5 & steps:* The point that indicates where we are on the map at the end of the 3rd level has disappeared.

- *Bug 6 & steps:* On the map of Australia where the levels are displayed, I can hear sound effects when moving the mouse over them on the first load. When I pass a round or go to the main menu and click CONTINUE, I cannot hear them (besides the music in the background). I cannot hear them even when choosing the color of the car.

- *Bug 7 & steps:* Using a different icons for the same functionality in the settings and at the level start (**Figure 8**).

Comment: Bugs appeared during testing on all test devices.

2.4.2 Testing of early-access game pillow bellow

This chapter presents an example of game testing of the early-access game carried out in 2021. The game is currently available on the Steam platform.
Game: Pillow Bellow.
Developer & Publisher: VOX Design.
Version: Early-Access.

Playtesters:

- The characteristics of playtesters are similar to the Bushfires game.

Playtesting:

- Positive remarks:
    - Very nice concept of the game with a unique world.
    - Refill life by collecting feathers.
    - I like that players can collect many different achievements in levels. For example, it is a lucky surprise to get the achievement of being lost in the woods.
    - I like it is possible to achieve also objects different from feathers, e.g., skins representing each game level.
    - It is exciting for a player that every level has a different theme, various traps, and enemies.
    - It is great that the way to the goal is not precisely defined. Mostly there are multiple paths, and to fulfill the level requirements, the player must explore all of them.

Figure 8.
Example of bushfires game – Different icons for the same functionality.

  - Using relaxation music which varies on all levels, there is little chance it will get on the player's nerves.

- Negative remarks:

  - The disadvantage is that there are no checkpoints in the levels, and it is always necessary to start over. It would be appropriate to add at least one checkpoint for each level.

  - The music at the Halloween level does not fit the theme. Something scary or indicating the Halloween theme would be more suitable.

- The third level represents the world of needles and threads. Sometimes the pillow gets stuck on a needle, and nothing happens. It would be good if the game reduced the player's health in such a case. If he does not jump off the needle at a defined time, he will die (like falling off the track).

- The third level represents the world of needles and threads. The player's pillow gets stuck in the object representing the threads, and it is challenging to move with it.

- Players cannot save collectibles (e.g., skins) and achievements in levels. Players cannot transfer skin (appearance) obtained in level 3 to other levels. It would be convenient to create a database of these objects and refer them to the main menu. The game would unlock the collectibles as the player found them in levels. The player would thus have an overview of his game activity. The player could "put on" the skin on his pillow whenever he wanted.

- In the Christmas special, the dialog with the villager switches by itself. I would allow the player to switch it so that he can read it. Not everyone reads quickly.

Usability Testing:

- I like the game has a tutorial, and right at the beginning, it nicely explains what to expect in each level and how to control the pillow.

- Turning off the game using the X key in the level map simplifies the game control.

- On some levels, places in the dark or the shadows are too dark, and it is difficult to orientate there.

- If impaired mobility on a thread-type object happens on purpose, I'd reduce the step length or make the pillow take three steps to move (shear effect).

- The tutorial will start first instead of the main menu when restarting the game. When starting the game, the main menu should appear first, where there should be a link to the tutorial and the menu of levels.

- The menu under options is shown at the bottom that one can go back only with the right mouse button, but it also works with Backspace. However, it is not marked anywhere.

Bugfixing (QA):

- *Bug 1 & steps:* Even though I turned off sounds and music, the sounds and music start at the beginning of each level. (Students attached video.)
- *Bug 2 & steps:* The counter of collected stars does not update in the Christmas special.
- *Bug 3 & steps:* In the Christmas special, I received a message about the successful completion of the level, although I did not reach the flag and did not even collect all the stars. The video shows that it was impossible to collect one of the stars. That is, I did not collect at least one star. (Students attached video.)
- *Bug 4 & steps:* In the level with the princess (level 5), my life got to zero after contact with the enemy, but even so, I was not sent back to the beginning and could continue the game. (Students attached video.)
- *Bug 5 & steps:* In the Halloween special, we need to collect 10 feathers, but the level counter shows we need to collect 15 (**Figure 9**).
- *Bug 6 & steps:* In the introductory tutorial and some level descriptions, there are typos.
- *Bug 7 & steps:* Moving objects sometimes disappear and appear in the levels. (Students attached video.)

2.4.3 Testing of student multiplayer FPS game NSD

This chapter presents an example of game testing of a multiplayer FPS game carried out in 2022 (**Figure 10**). A group of students developed this game as part of a bachelor's project.

Game: NSD.
Developer: Students.
Version: Early-Access.

Playtesters:

- The characteristics of playtesters are similar to the Bushfires game.

Figure 9.
Example of pillow bellow game – Bugs.

Playtesting:

- Positive remarks:
    - Since this is a multiplayer game, chat and private messages are a plus.
    - Spawning immortality makes spawn camping impossible.
    - A different map for each match increases the dynamics of the gaming experience.
    - Considering this is an FPS game, it is good that the game has enough obstacles on the map.
- Negative remarks:
    - Lack of feedback in the menu (friend request and invite).
    - Spawning too close to other players occasionally.
    - The possibility of "jump shots": A shot during a jump is accurate only if the player was standing before. So, it is possible to stand behind an obstacle, jump up, and shoot while the opponent is at a disadvantage.
    - After leaving the ladder, the player cannot shoot for a long time.
    - The shotgun is too weak (even in close combat).
- Other remarks:
    - The weapon movement based on the camera rotation looks interesting, but sometimes it is distracting.
    - One can use [Match] chat even if he is not currently in a match (and this chat corresponds to a previous match).

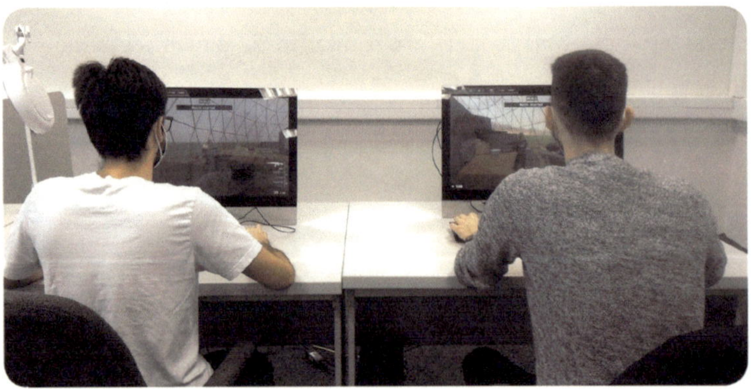

Figure 10.
Testing the multiplayer FPS game NSD in the classroom.

- One cannot leave the ongoing match.
- The character model (third-person view) does not match what the player sees in the first-person view. For example, the player sees the character is wearing gloves while holding the weapon/knife, but there are no gloves on the model.

Usability testing:

- The possibility of setting the aiming sensitivity.
- Ability to jump onto the roof of the building using the surrounding obstacles.
- After respawning, the player must select a weapon again (the player is empty-handed). With frequent respawn, this is a user disadvantage.
- Jumping through the windows is possible but difficult.
- SUS questionnaire:
  - Questions focused on movement, shooting, weapon damage tuning, and the overall UI experience.
  - 41 users completed it.
  - The Likert scale was 1–5.
  - Overall evaluation was 76.16, which is positive (B).

Bugfixing (QA):

- *Bug 1 & steps:* Getting stuck in a wall by closing a door (another player must close it; a person cannot close it himself if he is inside it).
- *Bug 2 & steps:* The possibility to stand on top of another player if he is crouching; in other cases, one can pass through players.
- *Bug 3 & steps:* The ability to jump on a tree and stay on it (according to the developers, it should not have been possible).
- *Bug 4 & steps:* If the character is holding a knife, when one quickly clicks the mouse button, the animation of using the knife starts over again, but the attack does not finish.

3. Conclusion

In our department, we generally focus on education in software engineering. So, in the design and development of games, the implementation side, including the iterative process, is vital to us. We are looking for ways and activities to teach students to transfer formalisms and critical aspects of Game Theory into practice. It is not only a transfer into source code but also a design or feedback.

This chapter presents our experience implementing the Design and Development of Computer Games and other similar courses. We emphasize several essential methods and aspects like iterative development, rapid (paper) prototyping, and game testing. In the chapter, we presented educational activities mainly focused on game testing: Iteration of a card game rules, design of a board game, design of a computer game with a paper prototype, and testing of early-access games. Activities are described in steps, including their motivation and benefits. They are supplemented with our commentary from the educational process point of view and students' practical outputs in both text and graphical form. The results include feedback in the testing forms: Focus Groups, Playtesting, Usability Testing, and Quality Assurance.

Critical thinking and the ability to transfer formalism into practice are crucial in the game industry and software engineering. Based on our experience, we can say that the students' outputs are mostly interestingly processed. They do not lack creativity or attention to detail, which we consider a positive result. The game creators processed early-access game testing. They are primarily experts working in the game industry who gave students feedback on how beneficial the testing was, being positive in most cases. Based on this, we perceive the results as contributional.

We believe that our work will inspire lecturers from similar fields. Given that the results presented in this chapter are based on multiple years of experience with students, they can also be used as a basis for further studies and they will fulfill the title in terms of the way from theory to practice.

Acknowledgements

This work has been supported by Slovak KEGA Agency under grant no. 002TUKE- 4/2021: "Implementation of Modern Methods and Education Forms in the Area of Cybersecurity towards Requirements of Labour Market" and under grant no. 048TUKE-4/2022: „Collaborative virtual reality technologies in the educational process".

Author details

Emília Pietriková and Branislav Sobota*
Technical University of Košice, Slovakia

*Address all correspondence to: branislav.sobota@tuke.sk

References

[1] Schnell J. The Art of Game Design: A Book of Lenses. Third ed. NY USA: CRC Press; 2019

[2] Fullerton T. Game Design Workshop: A Playcentric Approach to Creating Innovative Games. Fourth ed. NY, USA: CRC Press; 2018

[3] McBain J. Learning Game Design. Kindle ed. Melbourne. Australia: ASIN; 2016

[4] Choi JO et al. Playtesting with a Purpose. In: CHI PLAY '16: Proceedings of the 2016 Annual Symposium on Computer-Human Interaction in Play. NY, USA: 2016. pp. 254-265

[5] Lennart E et al. User experience (UX) research in games. In: CHI EA '19: Extended Abstracts of the 2019 CHI Conference on Human Factors in Computing Systems. NY, USA: 2019. pp. 1-4

[6] Medlock MC. The Rapid Iterative Test and Evaluation Method (RITE). First ed. San Francisco, USA: Oxford Press; 2018

[7] Mirza-Babaei P et al. A postmortem on playtesting: Exploring the impact of playtesting on the critical reception of video games. In: CHI '20: Proceedings of the 2020 CHI Conference on Human Factors in Computing Systems. NY, USA: 2020. pp. 1-12

[8] Roohi S et al. Predicting game difficulty and engagement using AI players. In: Proceedings of the ACM on Human-Computer Interaction. NY, USA: 2021. pp. 1-17

[9] Lewis J, Sauro J. Ten Things to Know about the RITE Method. Denver, USA: Measuring U; 2020

[10] Adams E. Fundamentals of Game Design. Third ed. NY, USA: New Riders; 2013

Chapter 8

Time Travel Gamification of Learning and Training: From Theoretical Concepts to Practical Applications

Klaus P. Jantke, Hans-Holger Wache and Ronny Franke

Abstract

Gamification is considered the systematic anticipation and design of affective experiences. It is not erroneously reduced to the usage of game-typical elements in another context. The human experiences in focus are varying forms of virtual time travel. In a time travel exploratory game, players return virtually to the past for gaining insights and, possibly, finding artifacts bring back to the present time. This works well for environmental education studying, by way of illustration, the world-wide ocean warming over several decades. Time travel prevention games go even further. Players who visit the past get an opportunity to impact their fate. This works well in application areas such as crime prevention and industrial accident prevention. Dynamic time travel prevention games are a recently developed game type in which the past changes dynamically to support the player's chances of successfully completing the mission. The authors present original concepts and technologies and demonstrate running applications.

Keywords: gamification, time travel games, time travel exploratory games, environmental education, time travel prevention games, industrial accident prevention, dynamic time travel games

1. Introduction

The Institution for Statutory Accident Insurance and Prevention for Raw Materials and Chemical Industry has more than 34,000 member enterprises in Germany with about 1.6 million insured persons. Learning and training that aims at accident prevention are of high societal significance.

Gamification is an approach to make conventional learning and training more attractive, even more effective. In particular, the authors present concepts of virtual time travel and implementations as well as applications.

2. Gameplay

The systematic transformation of learning and training into playful experiences requires an understanding of the phenomenon of play. Johan Huizinga's cultural-historical perspective [1] is worth consideration, but far from operationalization. Opinions are divided.

There are a few more concise conceptualizations in the topical literature such as considering *an activity taking place on the basis of formally defined rules and containing an evaluation of the efforts of the player. When playing a game, the rest of the world is ignored* [2]. However appealing at a first glance, this perfectly fits written exams in schools, vocational education, and higher education. It misses the point.

As Fabricatore puts it, *there is probably no universally accepted definition of gameplay* [3]. But there is an urgent need for a firm foundation of endeavors such as gamification in areas of societal significance.

Based on rich sources such as the books by Fritz [4] and Koster [5], the authors prefer a lucid approach adopted from [6, 7] and illustrated by means of **Figure 1**.

Those who engage in gameplay perform an act of framing as psychologists and communication scientists call it. That is basic to Juul's saying that the rest of the world is ignored (cited from reference [2] above). And it coincides with Bartle's saying that *at the persona level of immersion, the virtual world is just another place you might visit, like Sydney or Rome. Your avatar is simply the clothing you wear when you go there. There is no more vehicle, no more separate character. It's just you, in the world* (cited after [8]).

Engaged in playing a game, humans experience a balance of indetermination and self-determination [9]. This balance may depend on the knowledge and skills of the player and changes over time due to the experience of play. Indetermination may be based on a large variety of factors such as randomness, complexity—there is a deep relationship between randomness and complexity [10]—environmental conditions, and the presence of other human players. Apparently, because the balance is dynamic, the experience of play changes over time. What appeared incomprehensible in the beginning may become more and more interesting over time. What once has been exciting may become boring when getting under control.

The crux is that play brings with it learning—learning of the game mechanics, learning of patterns of play, learning of the relationship of stimulus and response, and the like. The player's efforts of gaining control to increase self-determination and to overcome indetermination unavoidably result in learning (for illustration, see [11]).

The authors' perspective at play visualized by means of **Figure 1** leads to some fundamental insights into the essentials of play such as the dovetailing of playing and learning. The perspective serves as a guideline for game design, in general, and for the gamification of learning and training, in particular. This will be demonstrated throughout the rest of the present contribution.

Figure 1.
Illustration of the concept of gameplay as introduced and discussed by the authors in [6, 7].

3. Gamification

The authors aim at the transformation of socially relevant learning and training into attractive and effective experiences of play, a case of educational gamification, a term coined in [12]. In this report, the authors cite a vision translated from [13] as follows. *Imagine—if successful—teaching, learning, and training offers that are addictive. Learners always want more and learn more, trainees cannot stop to train, and those who learn and train, form communities on the internet where they exchange their experiences and successes in acquiring knowledge and sharing skills, as well as inspiring others to participate. Imagine, textbook publishers producing educational materials of an addictive nature and where school becomes our children's favorite venue.*

The gamification of a publisher's textbook material is systematically investigated and technologically described in [14]. But this does not apply to the present chapter that has *Time Travel Gamification* in focus, a term coined here. There is apparent evidence for the need for disambiguation.

As Landers et al. put it, *definitions of gamification tend to vary by person, both in industry and within academia* ([15], p. 315).

Authors like Deterding et al. trivialize gamification as the use of elements of game design in non-game contexts ([16], p. 2, see also [17], p. 10). If they would be right, everyone were able to implement gamification and nobody would need any scientific background to do so. Chou [18] claims to go *beyond points, badges, and leaderboards*, but being a bit changeable, he later explains gamification as *the craft of deriving fun and engaging elements found typically in games* ([18], p. 8). Here they are again, Deterding's elements of game design.

Bogost [19] responds by calling gamification *bullshit* and naming those supporting gamification *bullshitters*.

At this point of the controversy, the authors take up a position. Due to the focus on learning and training, the term of educational gamification is adopted from [12], p. 5: *Educational Gamification means the transformation of given learning or training material and/or educational environment into a form that bears the potential of playful experiences that are likely to unfold when humans accept to engage.*

As briefly discussed in the preceding section, experiences of gameplay are highly individual and, in addition, depending on local and temporary conditions. According to Chou, gamification means *combining different game mechanics and techniques to form desired and joyful experiences for everyone* (cited from [18], p. 10).

There is no way "to form experiences for everyone". Instead, there are techniques of providing spaces of experience—called story spaces in [20]—such that there may dynamically unfold experiences of play that vary from one human player to the other, i.e. desired and joyful personalized experiences for everyone. Story space design is dynamic planning [21] based on dynamic approaches to storyboarding [22]. The storyboarding approach itself is adopted and adapted from dynamic plan generation in critical industrial application domains [23]. *Storyboarding is the organization of experience*, as Jantke and Knauf put it ([22], p. 25). To make the experience affective and effective, the design shall be adaptive to the learners' needs and desires [24–29] by means of Artificial Intelligence [12].

Varying approaches aim at transformations toward exciting playful experiences. In critical industrial application areas, failure is a crucial phenomenon to deal with. Following Litts and Ramirez, the authors see *failure as a process, not an endpoint* [30]. Papert compares the trainees' struggle with failure to a term of software engineering: debugging ([31], p. xiii). This is logically related to abductive reasoning [32, 33].

To deal with failure playfully, given a digital training environment, the authors suggest the implementation of abductive reasoning by means of virtual time travel, a novel approach to gamification called *time travel gamification*.

In other words, gamification of a training environment is performed by means of the introduction of time travel opportunities. Trainees get offered the chance to impact the fate—an exciting experience they can hardly gain in the industrial practice.

Imagine a training module structured according to the high-level storyboard graph on display in **Figure 2**. For more details of the underlying storyboard concept and of the storyboarding process, readers are directed, besides the key source [22], to a conceptualization called layered languages of ludology [34, 35], to dynamic plan generation [21, 36–42], to didactic knowledge in storyboards [43–48], and to large-scale applications in areas such as civil protection and disaster management [49, 50]. The details of the storyboarding technology are beyond the limits of this contribution.

To complete this short intermediate discussion, interested readers are directed to a possibly valuable practical approach—*storyboard interpretation technology* [51]. Sufficiently complete digital storyboards are computer programs that may be subject to automatic interpretation, i.e., to execution.

Smaller nodes of a storyboard graph describe *scenes*. Scenes have—possibly alternative—semantics in the domain given by other components of the storyboard. By way of illustration, the scene Task Formulation may be implemented by means of a video, an audio file, a PDF document, a text on the screen, or anything like this up to a face-to-face instruction by a trainer. The scene Exit Menu might offer just two buttons either for returning to the Data Analysis or for leaving the training session.

The larger nodes of a storyboard graph represent *episodes*. For every episode, there may exist alternative graphs of the storyboard for substitution. The substitution of a storyboard graph for a node has execution conditions, an issue that will be discussed in more detail below. If an execution condition contains variables that may hold dynamically changing values and if the substitution of nodes is postponed up to execution time, this is key to the dynamic emergence of varying training experiences, a decisive property inherited from the underlying approach to dynamic planning [23].

Gamification is the transformation of designs such as the one shown in **Figure 2** into the one in **Figure 3** toward the enabling of varying playful experiences.

The authors' present approach to time travel gamification consists of the extension of a given design by means of opportunities to travel back in time virtually. The didactic intention is to allow for another trial—perhaps several times—such that human trainees get a chance to arrive at better results, found in the Data Analysis episode, they may consider their own individual achievement.

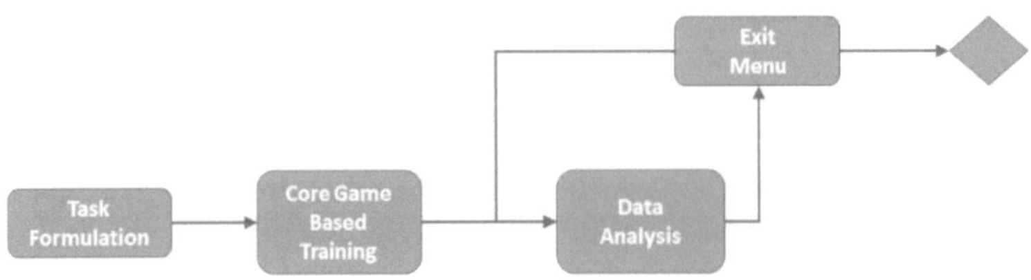

Figure 2.
Top-level design storyboard graph G1 specifying a training module including closing assessment.

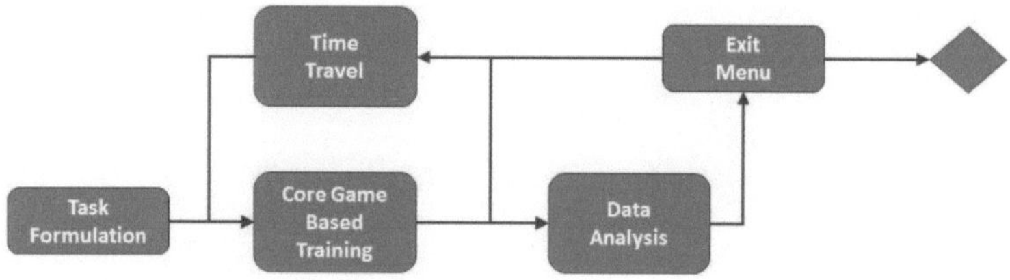

Figure 3.
Top-level design storyboard graph G1 specifying time travel gamification of a training module.

4. The time travel gamification methodology

The introduction of time travel does not primarily lead to any game elements such as points, badges, or leaderboards. It bears the potential of unprecedented experiences.

4.1 Syntactic embedding

Syntactically, time travel gamification shows in embedding novel scenes and episodes in conventional interaction design as exemplified by **Figure 3** compared to **Figure 2**. In the case of **Figure 3**, for instance, inserting a Time Travel episode requires changes in all implementations of the Exit Menu scene such that the novel episode becomes accessible from there. By way of illustration, **Figure 4** shows on the left a Tardis-like entrance to time travel. The present authors try to avoid overstating issues of syntax.

Similarly, the edges outgrowing newly introduced episodes and scenes need to point to some successor node such as core game-based training in **Figure 3**.

Under the hood, so to speak, the training system does a lot of bookkeeping to allow for adaptivity. User/learner/trainee/player profiles are essential to a digital system's ability to hypothesize the human user's needs and desires and to adapt accordingly. This artificial intelligence perspective is discussed in some detail in the paper [52]. The authors confine themselves to a detailed discussion of syntactic implicitness. From time to time—as in the following paragraphs—a few details are shining through.

Another more subtle issue of syntax relates to execution conditions. For the sake of adaptivity, it becomes important to process data of a trainee's repeated time travel. Those data are collected by means of the system's bookkeeping. They are only useful if execution conditions are syntactically modified to access the time travel history. It is an issue of didactics and game design on how to use the syntactic data semantically.

4.2 Time travel control

Time travel gamification has the gist of play in focus. Consequently, it comes with a high degree of self-determination that shows, among others, the freedom to select the destination of time travel. **Figure 4** shows on the right a variant of the authors' time tunnel as implemented in some of the authors' current applications.

During a training session, the episode of core game-based training is substituted by a lower-level storyboard graph, nodes of this graph may be substituted again, and so on until scenes determine what actions a player may execute. For executed scenes, the system uses iconic representations that represent destinations in the time tunnel. The

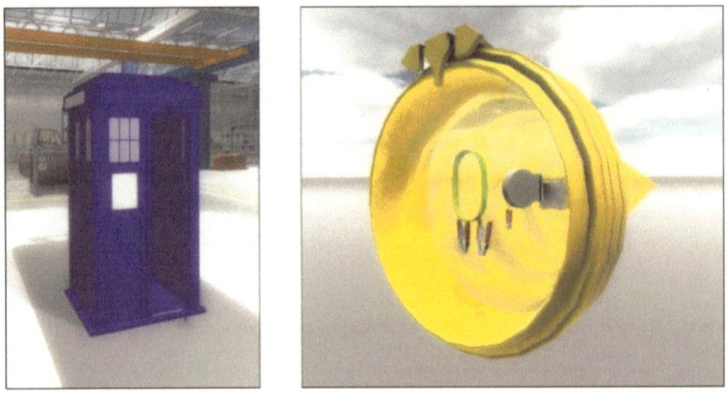

Figure 4.
Technicalities to control time travel: a Tardis-like entrance and a time tunnel with destinations.

player may fly through the time tunnel backward and forward in time, the latter—for the time being—only within the limits of the history of gameplay that took place. For time travel prevention games as in [21, 53], and for time travel exploratory games, a term coined recently in [52], journeys to the future are not yet considered. The selection of an object by clicking the button in the upper middle of the time tunnel brings the player backward in time to the related scene.

In case a player has difficulties in completing the training mission successfully and, therefore, goes repeatedly on a journey in time, this shows the necessity of guidance. The adaptivity of time travel is a key issue of time travel gamification. Consequently, time travel gamification is deeply interwoven with artificial intelligence design [52].

From the perspective of play (see **Figure 1**), the first approach to better adaptivity is to change the balance of self-determination and indetermination. The choice declines.

The first approach may be implemented by stepwise reduction of the stations in the time tunnel as demonstrated earlier by the authors (see [53], Figures 8 and 9). The ultimate strongest version is to direct the player to the decisive scene only. Apparently, it is an issue of learning psychology, game design, and media didactics whether or not and, if so, in what conditions to decide for such a strong guidance.

Similarly, one may reduce the number of possible interactions with a scene, in this way decreasing the likelihood of wrong decisions, a step toward dynamics of the past.

4.3 Dynamic modification of the game world

Dynamically changing the past of gameplay is a method introduced in [54] and demonstrated by means of an application for the training of accident prevention in the paint and coatings industry. Trainees when traveling back in time may arrive at scenes they have experienced before, but that appears somehow different this time.

The method is illustrated in the present contribution by means of two screenshots taken from the application demonstrated in the following Section 5.

There is an enormous space of opportunities to change the story [20] experienced during game-based training—a highly creative part of the time travel gamification methodology. Subsequent variations of AI support to the human trainee may be considered cascades, as the authors put it in the study [54].

Figure 5 shows on the left a scene as experienced for the first time. Among others, grounding cables and pumps are available. The trainee is expected to install the grounding before a pump is used. Furthermore, the wall-attached grounding shall be

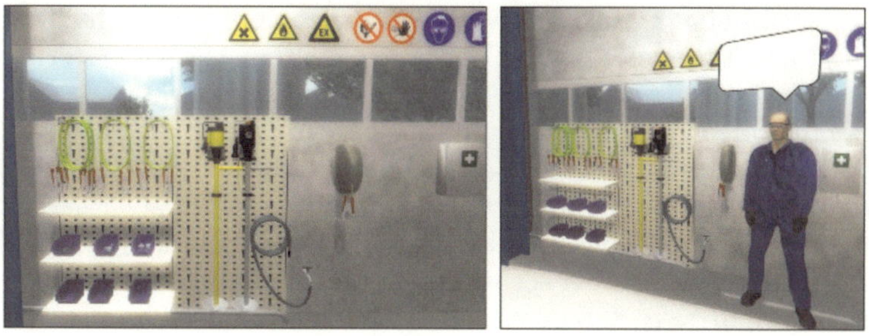

Figure 5.
The past is no longer what it used to be—dynamic changes to increase the likelihood of success.

used first, because attaching the two-sided grounding cable to two containers may result in an equipotential bonding.

At a later visit on display in **Figure 5** on the right, the virtual past has changed. There appears a nonplayer character (NPC) talking to the trainee. The empty speech bubble is intended to indicate that even the NPC's utterances may be modified from a visit to visit aiming at increasingly better guidance of the trainee.

5. The time travel gamification in practice

No other industrial training experience is lasting as long as a self-induced accident, especially if it comes with an explosion and fire. Affective training is more effective. For good reasons, however, the authors prefer to experience accidents only virtually. The installation on display in **Figure 6** is a perfect place for learning from experience, in particular, if trainees get offered opportunities of time travel to impact the fate.

The authors adopt and adapt some training situations from [53] in the virtual installation on display in **Figure 6** developed at Fraunhofer IFF, Magdeburg, Germany. The implementation is in use at the Institution for Statutory Accident Insurance and Prevention including several industrial partners. The mission discussed in the present demonstration consists of the decanting of an inflammable fluid from a larger container into a slightly smaller barrel. Trainees are informed about their mission in the task formulation scene (see **Figures 2** and **3**).

Figure 6.
A virtual installation for training accident prevention is subject to time travel gamification.

The left screenshot of **Figure 6** shows the virtual factory in its virtual environment. On the right, there is the inside workplace where the trainee shall perform the task explained by means of the task formulation scene implementation.

It follows the core game-based training episode that is implemented by several scenes of activities such as putting on protective clothing and turning on deaeration.

The scenes that occur in the main training episode are partially ordered. Some of them have a certain dependence on others—by way of illustration, getting dressed comes first—whereas some other scenes are mutually independent of each other. By way of illustration, there is no required ordering between turning on the de-aeration and attaching the two grounding cables. But for grounding, a required order exists.

The designer team of domain experts, IT specialists, learning psychologists, and possibly other experts negotiate those dependencies. The result is reflected in some storyboard graphs specifying potential implementations of the main episode [21].

Throughout this section, subsequent figures illustrate the actions of a trainee that took place within a training session. Recall that actions have iconic representations that occur in the time tunnel, in case traveling back in time becomes necessary. Subsequently, all this will be illustrated by means of screenshots from a training session that begins with the dressing scene (see left screenshot of **Figure 8**) and ends with an undesired accident visualized by the rightmost screenshot of **Figure 11**. Unfortunately, a book chapter presentation like the present one does not allow for a visualization of the dynamics of the human-system interaction.

In case an undesired event terminates the main training episode, a trainee has to decide how to respond. Time travel gamification is intended to reduce frustration, encourage continuation, and guide the trainee to a satisfying completion of mission that is experienced as the trainee's own achievement.

5.1 A straight forward training session

The designer team negotiates the issue and decides on the intended balance of self-determination and indetermination (see **Figure 1**). Potential substitutions of the core game-based training episode specify the dichotomy of freedom and guidance.

The simple storyboard graph on display in **Figure 7** exemplifies the design idea that getting dressed professionally first is not put at the human trainee's disposition. Trivialities like that are not subject to game-based training. This may serve just as an example, an illustration, of how storyboarding is used to adjust the focus of training.

Notice that the scene of the storyboard graph in **Figure 7** is indicated iconically. Seen from the perspective of time travel, scenes are considered atomic. Journeys may lead back to a scene as a whole, but not into a scene. This is underlying the iconic representations in the time tunnel (**Figure 4**).

Figure 8 reports the first trainee activities by means of 3 subsequent screenshots.

After putting on the personal protective equipment, the trainee dealt with a carriage blocking the workspace and, next, turned on the deaeration.

Figure 7.
*Next-level storyboard graph for expanding the core game-based training episode (**Figures** 2 and 3).*

Figure 8.
Preparatory scenes of training to get ready for the critical decanting of a certain inflammable fluid.

Notice that the carriage on display in the central screenshot of **Figure 8** carries a radio and a cellphone both containing rechargeable batteries. Those objects must be removed from a workplace; a sub-activity when dealing with the carriage scene.

There is a larger number of scenes—actions to be executed by the trainee—that have to occur in expansions of the game-based training episode shown in **Figure** 7. They are iconically visualized in **Table 1**.

In fact, **Table 1** represents a storyboard graph that exclusively consists of scenes. Whereas graph representations as in **Figures 2, 3**, and 7 might be more appropriate to human comprehension, a relational presentation such as in **Table 1** does perfectly meet a digital system's needs.

Conceptually, every finite directed graph consists of a set of nodes N and a set of edges E, where E is a binary relation over N, i.e. $E \subseteq N \times N$.

Following Jantke and Knauf [22], storyboard graphs are additionally annotated by a variety of logical formulas. Formulas that are attached to a whole graph determine conditions in which the graph may be used for expansion of an episode. By way of

	0	1	1	1	1	1	c_2	c_3	c_3	c_4
	1	0	1	1	1	1	c_2	c_3	c_3	c_4
	1	1	0	1	1	1	c_2	c_3	c_3	c_4
	1	1	1	0	1	1	c_2	c_3	c_3	c_4
	1	1	1	1	0	1	1	c_3	c_3	c_4
	1	1	1	1	1	0	1	c_3	c_3	c_4
	1	1	1	1	c_1	c_1	0	1	c_3	c_4
	1	1	1	1	0	0	0	0	c_3	c_4
	1	1	1	1	0	0	1	1	0	c_4

Table 1.
*Edges between scenes for expansion of the core game-based training episode of **Figures** 2 and 3.*

illustration, this type of execution conditions is used to prevent graphs from multiple substitution, in case an episode—due to virtual time travel—is visited repeatedly. Designer teams negotiate and specify the regulations of graph substitution based on their respective concepts of didactics, game design, learning psychology, and the like.

Furthermore, edges may carry annotations as well. If an edge from a node n_1 to a node n_2 is annotated by a logical formula c, the formula c determines in which conditions node n_1 may be immediately followed by node n_2.

The first graph in tabular form represents largely unconstrained opportunities of trainee activities. Trainees can even repeatedly return for playing a scene differently. Notice that this graph does not obey the design requirement exemplified in **Figure** 7.

Readers may easily recognize that the graph represented by means of **Table 1** consists of 10 nodes and 76 edges. Note that 0 describes the logical value *false*, thus, indicating that the corresponding edge does not exist. A graph representation as in **Figures 2, 3**, and 7 with 76 edges might appear a bit cumbersome.

Notice that a graph for substitution of **Figure** 7's game-based training episode results from the one in **Table 1** by elimination of the first row and the first column.

As said before, the above graph is the most unconstrained one that may occur in an early training phase—its logical execution condition is controlling in which conditions it may be substituted for the core game-based training episode in **Figures 2** and **3**.

The value 1 in a cell of the graph description table represents the logical value *true* indicating the scene denoted by the icon in the leftmost column may be always followed by the scene denoted by the icon in the uppermost row.

There are four nontrivial annotations to edges denoted by c_1, c_2, c_3, and c_4, resp., decorating 28 of the edges. The formulas describe minimal technological conditions.

Condition c_1 says that the object of the next scene to be visited, either the barrel or the container, must not yet be connected to the wall-mounted grounding cable. Formula c_2 says that, at least, one of the object's container and barrel must be in place. And c_3 is the requirement that both the container and the barrel are already in place. By way of illustration, this is an obviously necessary precondition for the installation of the two-sided grounding cable. Condition c_4 describes the requirement that all preparations are complete and the process of decanting may be started.

The icon heading the last rightmost column is showing the switch to turn on the pump for completing the mission. This scene has no successor scene in this graph.

So far, the trainee has experienced three scenes as illustrated in **Figure 8** which are the first scenes iconically represented in **Table 1**. The trainee has many choices to continue as determined by the third row of **Table 1**.

Training proceeds as on display in **Figure 9**. First, the trainee brings the barrel to the workplace. Next, the container is provided. Third, illustrated by the rightmost screenshot of **Figure 9**, the trainee inspects the available grounding cables and selects

Figure 9.
Further preparatory scenes of training to get ready for the critical decanting of a certain inflammable fluid.

the wall-mounted cable. As visible in the leftmost screenshot of **Figure 10**, this cable is attached followed by the two-sided grounding cable collecting container and barrel.

Condition c_3 is satisfied. This allows for the installation of a pump as determined in the last but one column in **Table 1**. The second and the third screenshots report that the trainee has installed a pump in the container and connected the pump's hosepipe to the barrel. Furthermore, the de-aeration iconically represented by the green button has turned on and the exhaust system is in place over the barrel.

The graph relationally represented in **Table 1** determines that the only remaining scene—the one iconically represented by the switch on top of the last table column—is the execution of the decanting process. Turning on the pump by pushing the button shown in the central screenshot of **Figure 11**, results in an explosion and fire.

5.2 Time travel gamification exemplified

The training activities surveyed and illustrated in the preceding subsection lead to an undesired event, an effect that is not uncommon in the training of complex tasks in risky conditions. From the viewpoint of learning psychology, events like that are valuable. Self-induced accidents are effective and bear the potential of effective learning and of sustainable insights and patterns of behavior. Moreover, they may be considered worth telling and, thus, contribute to the dissemination of the authors′ efforts in training for accident prevention in the industry.

But how to respond in the context of training? How to continue such a session? Time travel gamification is the authors′ answer to transform an undesired event into just the initially motivating experience within the framework of an adventurous experience of play that leads human trainees to mastery using their own resources. As shown in the light of **Figure 1**, self-determination shall overcome indetermination.

Figure 10.
A pump is installed, grounding cables are attached, and a pipe leads to the barrel.

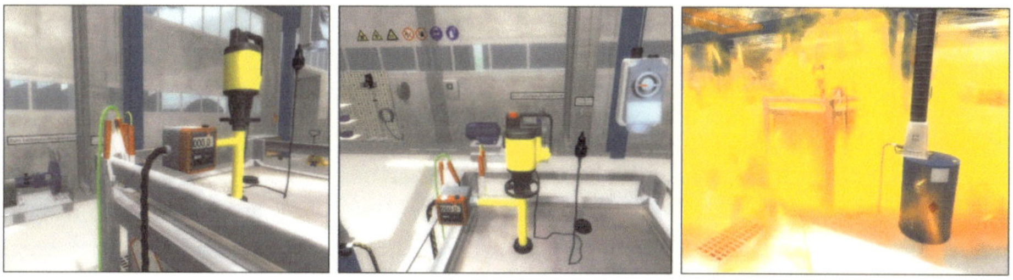

Figure 11.
The issue of pump installation in the container showing the counter, the switch, and the effect.

Digital storyboarding as reported in [21, 41–50, 52–54] is the design technology to span story spaces [20] of potential experiences desirable from a didactic perspective.

According to [22] expanding on [23], a storyboard is a finite hierarchically structured family of finite directed graphs, a bit more precisely, of pin graphs [55]. Lower-level storyboard graphs are prepared expansions for episodes in higher-level graphs. Among others, they represent patterns of game design and didactics [7, 47]. The modeling of educational theory by means of storyboarding is exemplified in [48]. In this section, the authors confine themselves to the present application case study.

First of all, time travel gamification means extending a storyboard by graphs such as the one on display in **Figure 3**. Trainees get opportunities to revisit scenes of the past—based on the system's bookkeeping of game-playing history—and to do better this time. But the avoidance of inaccuracies is not guaranteed by repetition of actions.

The diagonal structure of **Table 2** reflects a certain partition of the set of scenes.

The scene of putting on the personal protective equipment must come first because there is no edge in the graph that leads to this scene. This is followed by the three scenes dealing with the carriage, the leakage pump, and the de-aeration switch. These three scenes may be visited in any order and they may be revisited repeatedly. They form, so to speak, an elementary preparation episode.

The elementary preparation is followed by bringing the container and the barrel into the right working position. The corresponding novel execution condition c_5 says that the elementary preparation must be complete. Condition c_3 is the same as before saying that both the container and the barrel must be in place, whereas the negation written as $\neg c_3$ says that one of them is still missing. The wall-mounted grounding is usable as soon as c_3 is valid. The other scenes follow deterministically.

	0	1	1	1	0	0	0	0	0	0
	0	0	1	1	c_5	c_5	0	0	0	0
	0	1	0	1	c_5	c_5	0	0	0	0
	0	1	1	0	c_5	c_5	0	0	0	0
	0	0	0	0	0	$\neg c_3$	c_3	0	0	0
	0	0	0	0	$\neg c_3$	0	c_3	0	0	0
	0	0	0	0	0	0	0	1	0	0
	0	0	0	0	0	0	0	0	1	0
	0	0	0	0	0	0	0	0	0	1

Table 2.
Alternative graph implementing some concepts of didactics and game design for later expansion of the core game-based training episode when being revisited due to the chance of time travel.

Table 2 defines a storyboard graph more constrained than the graph in **Table 1**. The team of designers, in response to varying trainee behaviors, may come up with further alternative graphs aiming at a guidance to ultimately successful experiences. In this contribution, the authors refrain from an attempt of completeness. Instead, the graph of **Table 2** will be rewritten, visualized differently, and discussed in some detail.

The graph of **Table 2** presents an alternative characterized by a unique initial scene and a unique final one. Those distinguished nodes of a graph are called *pins* [56] expanding on [55]. Graphs like the one represented by **Table 2** and visualized differently by means of **Figure 12** are, so to speak, prototypical pin graphs.

From the viewpoints of ludology, of learning psychology, and of topical didactics, the goal is to support the emergence of varying experiences. This is made possible by means of the dynamics of plan generation as developed in [23] and adopted by [22]. Graph expansion occurs at execution time. The anticipation of varying experiences and the design of graphs prepared for substitution that—by means of time travel—make desired experiences likely is what the authors call *time travel gamification*.

By way of illustration, the core game-based training episode in **Figure 3** may be substituted differently when arriving there on repeated occasions. An expansion by means of the graph of **Table 2** leads to a closer look at time travel opportunities based on the structure that becomes more intuitively visible in **Figure 12**.

Figure 13 displays a design decision from the time travel gamification process that consists in offering to the human trainee 4 destinations in the time tunnel (**Figure 4**). The icon shown in the time tunnel represents the first scene of the related episode that occurred in the history of gameplay.

Note that the dotted arrows in **Figure 13** (and in **Figures 14** and **15**) are intended to represent the embedding of the excerpt on display into the graph of **Figure 3**.

For combinatorial reasons, there exist 14 different slightly more constrained variants of the (excerpt of the) graph in **Figure 13**. It is a design decision within the process of time travel gamification which of these graphs to include in the storyboard and what execution condition to attach to a graph.

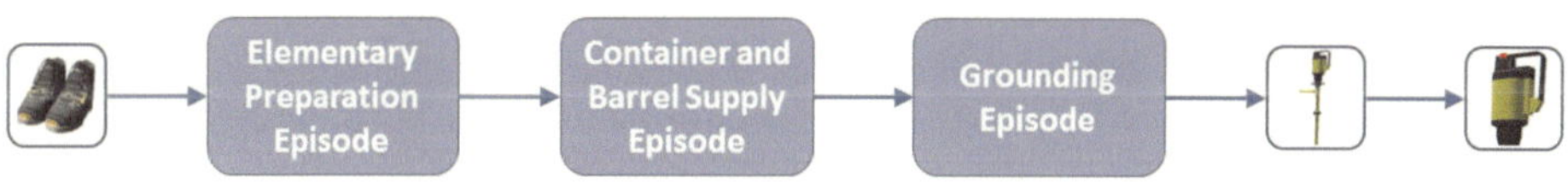

Figure 12.
*The prototypical pin graph of **Table 2** for expansion of the core game-based training episode.*

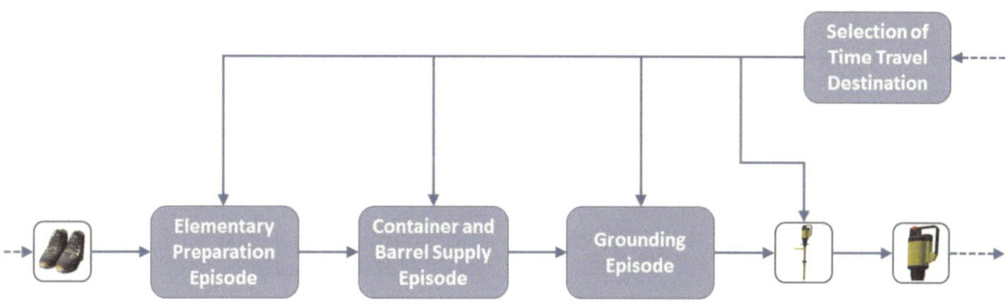

Figure 13.
*Excerpt of a storyboard after an expansion of the time travel episode as on display in **Figure 3**.*

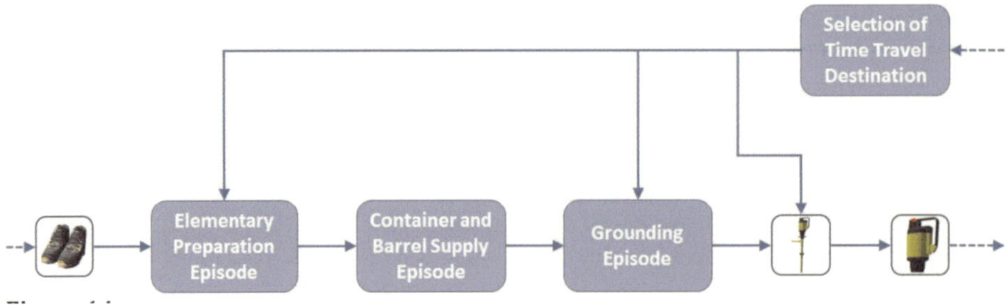

Figure 14.
An excerpt alternative to the one in ***Figure 13****; the present one being slightly more restrictive.*

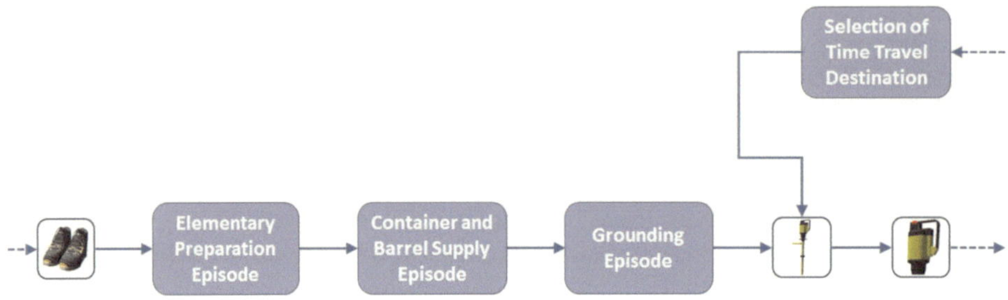

Figure 15.
Most restrictive time travel offer directing the trainee's attention to the cause of the accident.

An inspection of the training session surveyed in Section 5.1 reveals the cause of the accident illustrated by means the rightmost screenshot of **Figure 11**—the usage of an illegitimate pump. In case a trainee repeats this failure several times, a strict guidance by means of a very restricted time travel offer as on display in **Figure 15** may be the only way out.

Furthermore, the authors have developed the concept of dynamic time travel prevention games [54] in which even the virtual past is no longer what it used to be. The right screenshot of **Figure 5** illustrates a dynamically changing scene in which an NPC assists the trainee in selecting the right pump and, thus, completing the mission (**Figure 16**).

The ultimate goal of time travel gamification is the trainee's individual success.

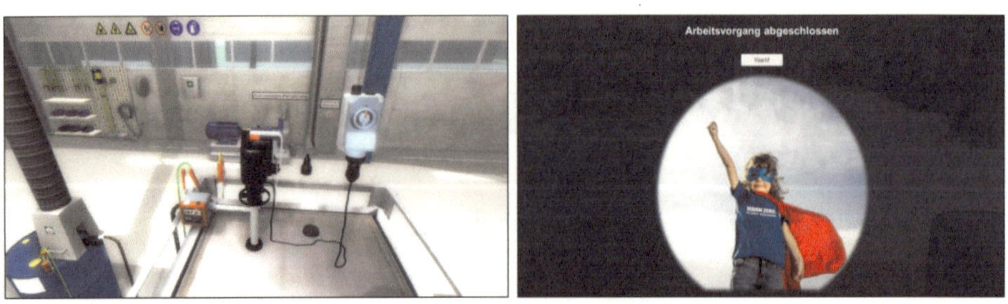

Figure 16.
The correct installation of all details including the selection of the right pump leads to success; illustration on the right © Shutterstock/sunny studio.

5.3 Technicalities of time travel gamification

The authors′ present chapter on time travel gamification is intended to be an introductory one spanning the spectrum of issues discussed from theory to practice dealing with original theoretical concepts and with practical industrial applications. Clearly, there are a large number of technicalities that are mostly kept under the hood. The present subsection, for the sake of honesty and correctness, addresses just a few.

According to [22, 23, 36], storyboards are finite hierarchically structured families of finite directed pin graphs. Computerized graph manipulation is tedious [36, 55, 56]. In the authors′ approach, the key technicalities are graph rewriting by means of node expansion underlying what the author calls storyboard interpretation technology [51]. The power of the approach, inherited from [23], lies in reasoning at execution time. When training begins, there does not yet exist the full graph of opportunities to come. Instead, this graph unfolds, so to speak, during gameplay and, thus, emerges in dependence on the trainee's individual behavior. Thus, experiences are personalized.

Intuitively, there is a top-level graph G_1 in the storyboard F such as in **Figure 3**. The interaction begins at an entry pin of this top-level graph G_1. When interpretation arrives at a scene, its operational semantics is used according to the currently valid execution conditions. Notice that issues of conflict have to be resolved. This holds as well when in a graph G episode e is reached. A graph $G_k \in F$ with a valid execution condition replaces this episode node e. A formal notation of expansion is $G[e \downarrow G_k]$. In this way, the current graph is rewritten to $G' = G[e \downarrow G_k]$. This establishes a rewrite relation $\Rightarrow_F$ in dependence on the storyboard F. At the end of gameplay, the initial graph G_1 is transformed into a graph G^m describing what ultimately took place. Formally, $G_1 \Rightarrow {}^*_F G^m$ where $\Rightarrow^*_F$ denotes the reflexive, transitive closure of $\Rightarrow_F$.

Originally, this is an approach to plan generation based on usually incomplete information in complex environments [23, 36]. Planning algorithms that implement graph rewriting are carried over to storyboarding (see [21], page 7, **Figure 4**).

Kirsten demonstrates formally that this dynamic plan generation approach is more expressive than alternative formalisms [57]. Essentially, the approach exceeds the limits of context-free formal languages (see [58] or any other standard reference). To circumscribe this formal result very loosely, with the planning approach of [23] one can generate a considerably larger variety of plans than with other approaches. Carried over to storyboarding, with the approach inherited from [23] demonstrated in [21, 49–54], one can generate a particularly rich space of potential experiences varying from trainee to trainee and depending on dynamic conditions of gameplay.

For several details and their practical relevance, interested readers are directed to [51] where the term storyboard interpretation technology is coined. Intuitively, this may be considered a technology of graph rewriting at progressing execution time.

Within this contribution, just one technical issue shall be discussed in more detail. For the sake of easy access, earlier examples and illustrations are revisited.

Graph rewriting as above works well even with cyclic storyboard graphs. But in the process of time travel gamification, certain cycles are of key relevance. Cycles representing phenomena of time travel do not aim at the repetition of actions that are the same as before, but at modifications that bear the potential of changes. Those graphs need a particular treatment that requires an extension of rewriting beyond the limits of the original approach [23, 36]. Time travel gamification provides feedback to the theory. To this end, earlier notions and notations are revisited [59].

From the perspective of the rewrite process, the goal is to remember the earlier steps of rewriting. From the perspective of technicalities, the methodology is a certain dotted notation (see [59], Section 3.2, definition 4 and **Figure 5**). From the perspective of experience design, the key is to enable the nonmonotonicity of story evolution [20]. That is the gist of time travel to impact the fate.

From a logical perspective, modalities such as possibility and necessity [60–62] play a fundamental role. Trainees who rely on time travel to fix problems of the present time depend on opportunities to change the past. Varying variants of the past are like Kripke models of modal logic. For the basics, readers are directed to [60, 61]. Furthermore, [62] provides a considerably large number of insightful discussions. Varying pasts of gameplay are possible, so to speak, to happen in the future of play. Loosely speaking, varying pasts seen as Kripke models are different parallel worlds that are visited one after the other during gameplay trying to change for the better. Subsequently, a few notational technicalities will be supplemented for clarification. In this way, the rewrite relation $\Rightarrow_F$ will be slightly modified.

When a node e is substituted by some graph G_k, all the nodes in $G[e \hookleftarrow G_k]$ that result from the inserted graph G_k are renamed. If n is such a node in G_k, [59] renames the new node to *e.n*. Apparently, this dotted notation helps to remember which expansion introduced this novel node to the resulting graph. In time travel gamification, the authors go even further. If episode e is reached for the first time, inserted nodes n of a storyboard graph G_k are renamed to *e.1.n*. If due to time travel, episode e is reached for the second time, a node n newly inserted is renamed to *e.2.n*, and so on. This dotted notation indicates on which occasion a node has been inserted.

Alternative pasts of play are represented by subgraphs that are mutually different. They form universes that reside literally in parallel within the history of interaction.

A firm background and technicalities like those sketched above back up sayings such as *the past is no longer what it used to be* and, to repeat a formulation from above, *varying pasts are possible to happen in the future*. The foundations provide meaning.

Notice that Papert's debugging aspect—seen as a technicality in the small—cited above according to [31] leads to a perspective from which the overall training process is seen as a sequence of actions and interactions, like the execution of a program. Traveling back in time allows for *patching the past*. Patching may be considered insufficient due to its potential renunciation of a global understanding [63]. This is an issue of bounded rationality separating "the beliefs that people have and the choices they make from the optimal beliefs and choices" (see [64], p. 449). The digital system must guide human trainees to ultimate success contributing to sustainable training effects—the artificial intelligence perspective at time travel [52]. The cognitive science issue of *fast and frugal heuristics* [65, 66] is related but exceeds the limits of the present contribution.

A few closing remarks are intended to illustrate the interplay of technicalities. Modalities of possibility and necessity as mentioned above—by way of illustration, the

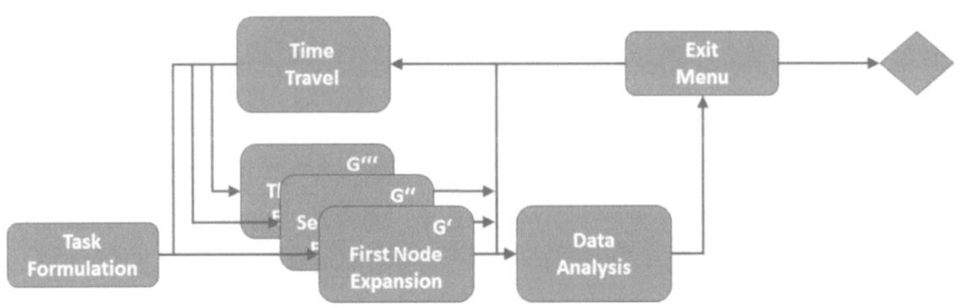

Figure 17.
Variants of the past occur subsequently establishing models of the world existing in parallel.

possibility of an accident—are controlled by incidences within storyboard graphs specified in incidence matrices like those on display in **Tables 1** and **2**. The storyboard, as a whole, specifies the space of conceivable experiences of which some unfold throughout gameplay whereas others do not. As illustrated by means of **Figure 17**, some are "conceivable but not actual" ([67], see also [62], p. 19). Designers negotiate logical conditions that occur in the incidence matrices to make desirable states likely.

6. Conclusions

Industrial accident prevention training draws benefits from time travel gamification that leads to practically applicable time travel prevention games [13, 21, 53, 54]. Similarly, environmental education benefits from time travel exploratory games [52].

However pleasant the state of science, technology, and applications might appear, this should not blind us to deficiencies and open problems. To mention just one manifest issue, the authors did not yet undertake any effort to travel into the future—virtually, in training games, an aspect already briefly investigated in [53], Section 3. Time travel gamification that includes the design of journeys to the future depends on ideas, concepts, and technicalities beyond those investigated in this contribution. By way of illustration, how to represent destinations in a tool such as our time tunnel? A scenery not seen before can hardly be represented iconically by objects to come. Further open problems arise when thinking of and discussing the design of the future. In the authors' present approach, the past serves as a blueprint for variants of the past that may deviate from earlier variants in cascades as the authors put it in [54] and as discussed in Section 4.3. The crux of time travel to the future is that this future is less constrained—a challenge to further work on time travel gamification.

Acknowledgements

Part of the work reported here including the present publication has been supported by the German Federal Ministry of Labor and Social Affairs (BMAS) with a prize awarded for the authors' concept entitled "AI on the Fly" which aims at the low-threshold communication of human-centered AI to a wide audience.

The authors gratefully acknowledge the cooperation with Winfried Wehrstedt who also belongs to the prize-winning team.

Last but not least, they gratefully acknowledge the fruitful cooperation with Oksana Arnold that has found expression in publications such as [21, 52–54]. She is, so to speak, the mother in spirit of dynamic plan generation [23] underlying the approach to storyboarding that is powering didactic and game design.

Author details

Klaus P. Jantke[1]*, Hans-Holger Wache[2] and Ronny Franke[3]

1 ADICOM Software, Weimar, Germany

2 Institution for Statutory Accident Insurance and Prevention for Raw Materials and Chemical Industry, Prevention Center Berlin, Germany

3 Fraunhofer Institute for Factory Operation and Automation IFF, Magdeburg, Germany

*Address all correspondence to: klaus.p.jantke@adicom-group.de

References

[1] Huizinga J. Homo ludens: Vom Ursprung der Kultur im Spiel. Amsterdam: 1939

[2] Juul J. A clash between game and narrative [PhD thesis]. Denmark: University of Copenhagen; 1999

[3] Fabricatore C. Gameplay and game mechanics design: A key to quality in videogames. In: OECD Expert Meeting on Videogames and Education. Paris: OECD; October 29-31, 2007; Santiago de Chile, Chile

[4] Fritz J. Das Spiel verstehen: Eine Einführung in Theorie und Bedeutung. Weinheim & München, Germany: Juventa; 2004

[5] Koster R. A Theory of Fun for Game Design. Scottsdale, AZ, USA: Paraglyph Press; 2005

[6] Jantke KP. Digital game knowledge media. In: Proceedings of the 3rd International Symposium on Ubiquitous Knowledge Network Environment, February 27–March 1, 2006. Sapporo Convention Center, Sapporo, Japan, Volume of Keynote Speaker Presentations, Sapporo, Japan: Hokkaido Univ; 2006. pp. 53-83

[7] Jantke KP. Pattern concepts for digital games research. In: Jantke KP, Kreuzberger G, editors. Knowledge Media Technologies: First International Core-to-Core Workshop, July 2006. Ilmenau, Germany: Technical University. pp. 29-40

[8] Wallace M. In celebration of the inner rouge. The Escapist. 2006;**30**:3-6

[9] Deci E, Ryan R. Handbook of Self-Determination Research. Rochester, NY, USA: Rochester University Press; 2004

[10] Schnorr CP. Zufälligkeit und Wahrscheinlichkeit: Eine algorithmische Begründung der Wahrscheinlichkeitstheorie. Berlin, Heidelberg, New York: Springer; 1971

[11] Cash B, Souki F, Games S. Oops, I learned something: Teaching via game mechanics. In: Martin C, Ochsner A, Squire K, editors. Proceedings GLS 8.0: Games + Learning + Society Conference, Madison, WI, USA, June 13-15, 2012. Pittsburgh: ETC Press; 2012. pp. 73-80

[12] Arnold O, Jantke KP. Educational Gamification & Artificial Intelligence. ADICOM Tech Report, 03-2018, Version 1.00. Weimar, Germany: ADICOM Software KG; 2018

[13] Jantke KP. No Thrill – No Skill: Ein systematischer Zugang zum Konzept Gamification. In: ADICOM Tech Report, 02-2018, Version 1.00. Weimar, Germany: ADICOM Software KG; 2018

[14] Friedemann S, Baumbach L, Jantke KP. Textbook gamification: Transforming exercises into playful quests by using webble technology. In: Helfert M, Restivo MT, Uhomoibhi J, Zvacek S, editors. Proc. 7th Intl. Conf. on Computer Supported Education. Lisbon, Portugal: CSEDU; 2015. pp. 116-126

[15] Landers RN, Auer EM, Collmus AB, Armstrong MB. Gamification science, its history and future: Definitions and a research agenda. Simulation & Gaming. 2018;**49**:315-337

[16] Deterding S, Sicart M, Nacke LE, O'Hara K, Dixon D. Gamification: Using game design elements in non-gaming contexts. In: Proceedings CHI´11, Extended Abstracts on Human Factors in

Computing Systems, May 7-12, 2011, New York: ACM; 2011. pp. 2425-2428

[17] Deterding S, Dixon D, Khaled R, Nacke LE. From game design elements to gamefulness: Defining "gamification". In: MindTrek '11: Proceedings of the 15th International Academic MindTrek Conference: Envisioning Future Media Environments, September 28-30, 2011, New York: ACM; 2011. pp. 9-15

[18] Chou Y-K. Actionable Gamification: Beyond Points, Badges, and Leaderboards. Fremont, CA, USA: Octalysis Media; 2016

[19] Bogost I. Gamification is bullshit. The Atlantic. August 8, 2011

[20] Jantke KP. The evolution of story spaces of digital games beyond the limits of linearity and monotonicity. In: Iurgel IA, Zagalo N, Petta P, editors. Proceedings of the 2nd International Conference on Digital Storytelling 2009, LNCS. Vol. 5915. Berlin, Heidelberg: Springer; 2009. pp. 308-311

[21] Arnold O, Franke R, Jantke KP, Wache H-H. Dynamic plan generation and digital storyboarding for the professional training of accident prevention with time travel games. In: Guralnick D, Auer ME, Poce A, editors. Innovations in Learning and Technology for the Workplace and Higher Education. TLIC 2021, Lecture Notes in Networks and Systems. Vol. 349. Cham: Springer Nature; 2022. pp. 3-18

[22] Jantke KP, Knauf R. Didactic design through storyboarding: Standard concepts for standard tools. In: Baltes BR, Edwards L, Galindo F, Hvorecky J, Jantke KP, Jololian L, et al editors. Proceedings of the 4th International Symposium on Information and Communication Technologies, Cape Town, South Africa, January 3–6, 2005. Dublin, Ireland: Computer Science Press; 2005. pp. 20-25.

[23] Arnold O. Die Therapiesteuerungskomponente einer wissensbasierten Systemarchitektur für Aufgaben der Prozeßführung. In: DISKI. St. Augustin. Vol. 130. Germany: Infix; 1996

[24] Thorndike EL. Individuality. Boston, MA, USA: Mifflin; 1911

[25] Dewey J. The child and the curriculum. In: Archambault AD, editor. John Dewey on Education: Selected Writings. New York, NY, USA: Modern Library; 1964

[26] Cronbach LJ, Snow RE. Aptitudes and Instructional Methods: A Handbook for Research on Interactions. New York, NY, USA: Irvingston; 1977

[27] Corno L, Snow RE. Adapting teaching to individual differences among learners. In: Wittrock MC, editor. Handbook of Research on Teaching. New York, NY, USA: Macmillan; 1986. pp. 605-629

[28] Tobias S. Another look at research on the adaptation of instruction to student characteristics. Educational Psychologist. 1989;**24**:213-227

[29] Denham A. Let's talk about intelligent tutoring systems and games for learning. In: Martin C, Ochsner A, Squire K, editors. Proceedings GLS 8.0: Games + Learning + Society Conference, Madison, WI, USA, June 13-15, 2012. Pittsburgh: ETC Press; 2012. pp. 5-8

[30] Litts BK, Ramirez D. Making people fail: Failing to learn through games and making. In: Ochsner A, Dietmeier J, Williams CC, Steinkuehler C, editors.

Proceedings GLS 10: Games + Learning + Society Conference, Madison, WI, USA, June 10–13, 2014. Pittsburgh: ETC Press; 2014. pp. 160-166

[31] Papert S. Mindstorms: Children, Computers and Powerful Ideas. New York, NY, USA: Basic Books, Inc.; 1980

[32] Peirce CS. A syllabus of certain topics in logic. In: Pierce Edition Project, editor. The Essential Peirce: Selected Philosophical Writings 1893–1913. Vol. 2. Bloomington, IN, USA: Indiana University Press; 1998

[33] Josephson JR, Josephson SG, editors. Abductive Inference: Computation, Philosophy, Technology. Cambridge, UK, and New York, NY, USA: Cambridge University Press; 1994

[34] Jantke KP. Layered Languages of Ludology: The Core Approach. Diskussionsbeiträge 25, Ilmenau, Germany: Techn. Univ. Ilmenau, IfMK; 2006

[35] Lenerz C. Layered languages of ludology – Eine Fallstudie. In: Beyer A, Kreuzberger G, editors. Digitale Spiele – Herausforderung und Chance: Beiträge der Tagungen LIT 2006 und 2007. Boizenburg, Germany: Verlag Werner Hülsbusch; 2009. pp. 39-52

[36] Arnold O, Jantke KP. Therapy Plan Generation in Complex Dynamic Environments. ICSI Report TR-94-054. Berkeley, CA, USA: International Computer Science Institute; 1994

[37] Arnold O, Jantke KP. Therapy plans as hierarchically structured graphs. In: Fifth International Workshop on Graph Grammars and their Application to Computer Science, Williamsburg, VA, USA. 1994

[38] Arnold O, Jantke KP. Variants of plan generation for complex dynamic systems. In: Rao X, editor. Proc. 8th Australian Conference on Artificial Intelligence, November 13-17, 1995, Canberra, Australia. World Scientific Publ. Co.; 1995. pp. 531-538

[39] Arnold O, Jantke KP. Related monotonicity phenomena in complex dynamic processes and in plan generation and simulation. In: Biundo S, Dean T, Waldinger R, editors. 12th European Conference on Artifcial Intelligence, ECAI'96 Workshop on Cross-Fertilization in Planning, August 12, 1996, Budapest, Hungary. European Coordinating Committee for Artificial Intelligence (ECCAI); 1996. pp. 41-48

[40] Jantke KP, Arnold O, Lehmann T. Flexible simulation scenarios for real-time planning in dynamic environments. In: Stewman JH, editor, Proc. Florida AI Research Symposium (FLAIRS-96), Special Track on Real-Time Planning and Reacting, Key West, FL, USA, May 20-22, 1996. Florida AI Research Society. Menlo Park: AAAI; 1996. pp. 90-95

[41] Arnold O, Jantke KP. Inductive program synthesis for therapy plan generation. New Generation Computing. 1997;**15**:27-58

[42] Arnold O, Jantke KP. Monotonic plan generation assistance: Challenges, open problems & solutions. In: International Conference on Systems and Informatics (ICSAI 2012), May 19-20, 2012, Yantai, Shandong, Piscataway: IEEE; 2012. pp. 215-220

[43] Jantke KP. Managing Didactic Knowledge in e-Learning: Storyboarding and Dramaturgy. Forschungsbericht 2004–01. Leipzig, Germany: Forschungsinstitut für Informations Technologien Leipzig e.V; 2004

[44] Knauf R, Jantke KP. Storyboarding—An AI technology to

represent, process, evaluate, and refine didactic knowledge. In: Jantke KP, Kreuzberger G, editors. Knowledge Media Technologies: First International Core-to-Core Workshop, July 2006. Ilmenau, Germany, Technical University. pp. 170-179

[45] Jantke KP, Knauf R, Gonzalez AJ. Storyboarding for playful learning. In: Reeves T, Yamashita S, editors. ELearn 2006, World Conference on E-Learning in Corporate, Government, Healthcare & Higher Education, Honolulu, HI, USA, Oct. 17-20, 2006. Chesapeake, VA: AACE; 2006. pp. 3174-3182

[46] Knauf R, Sakurai Y, Tsuruta S, Jantke KP. Modeling didactic knowledge by storyboarding. Journal of Educational Computing Research. 2010;**42**:355-383

[47] Jantke KP. Pedagogical patterns and didactic memes for memetic design by educational storyboarding. In: Arnold O, Spickermann W, Spyratos N, Tanaka Y, editors. Webble Technology, First Webble World Summit, WWS 2013, Erfurt, Germany, June 2013, Communications in Computer and Information Science. Vol. 372. Heidelberg: Springer; 2013. pp. 143-154

[48] Krebs J, Jantke KP. Methods and technologies for wrapping educational theory into serious games. In: Zvacek S, Restivo MT, Uhomoibhi J, Helfert M, editors. Proceedings of the 6th International Conference on Computer Supported Education, CSEDU 2014, Barcelona, Spain, May 1-3, 2014. Setubal: SCITEPRESS; 2014. pp. 497-502

[49] Arnold S, Fujima J, Jantke KP. Storyboarding serious games for large-scale training applications. In: Foley O, Restivo MT, Uhomoibhi J, Helfert M, editors. Proceedings of the 5th International Conference on Computer Supported Education, CSEDU 2013, Aachen, Germany, May 6-8, 2013. Setubal: SCITEPRESS; 2013. pp. 651-655

[50] Arnold S, Fujima J, Jantke KP, Karsten A, Simeit H. Game-based training for executive staff of professional disaster management: Storyboarding adaptivity of game play. In: Tan D, editor. Proceedings of the International Conference on Advanced Information and Communication Technology for Education (ICAICTE 2013), Sept. 20-22, 2013, Hainan, China. Atlantis Press; 2013. pp. 68-73

[51] Fujima J, Jantke KP, Arnold S. Digital game playing as storyboard interpretation. In: Proceedings of the 5th International Games Innovation Conference (IGIC), Vancouver, BC, Canada, September 23-25, 2013. Piscataway: IEEE Consumer Electronics Society; 2013. pp. 64-71

[52] Arnold O, Jantke KP. The Time Travel Exploratory Games Approach: An Artificial Intelligence Perspective. In: Csapó B, Uhoboimbhi J, editors. Computer Supported Education, 13th International Conference, CSEDU 2021, Virtual Event, April 23–25, 2021, Revised Selected Papers, CCIS. Vol. 1624. Heidelberg: Springer; 2022. pp. 40-54

[53] Arnold O, Franke R, Jantke KP, Wache H-H. Professional training for industrial accident prevention with time travel games. International Journal of Advanced Corporate Learning. 2022;**15**: 20-34

[54] Arnold O, Franke R, Jantke KP, Wache H-H. Cascades of concepts of virtual time travel games for the training of industrial accident prevention. In: The Learning Ideas Conference 2022, Proceedings [in Print]

[55] Höfting F, Lengauer T, Wanke E. Processing of hierarchically defined

graphs and graph families. In: Data Structures and Efficient Algorithms, LNCS. Vol. 594. Heidelberg: Springer; 1992. pp. 44-69

[56] Lohrey M. Model-checking hierarchical structures. Journal of Computer and System Sciences. 2012;**78**: 461-490

[57] Kirsten D. Properties of formal languages of therapy plans created by graph grammars. In: Kutrib M, Worsch T, editors. 5. Theorietag "Automaten und Formale Sprachen", Rauischholzhausen, Germany. Giessen: Justus-Liebig-University; 1995. pp. 132-142

[58] Salomaa AK. Formal Languages. New York, NY, USA: Academic Press; 1973

[59] Jantke KP. Developing and investigating two-level grammar concepts for design. In: Communications of the Algorithmic Learning Group, CALG 03/95. Leipzig, Germany: HTWK; 1995

[60] Feys R. Modal Logics. Amsterdam, Netherlands: E. Nauwelaerts; 1965

[61] Chagrov A, Zakharyaschev M. Modal Logic. Oxford University Press; 1997

[62] Goldblatt R. Mathematical modal logic: A view of its evolution. In: Gabbay DM, Woods J, editors. Handbook of the History of Logic. Vol. 7. Elsevier BV; 2006

[63] Ginat D. Hasty design, futile patching and the elaboration of rigor. ACM SIGCSE Bulletin. 2007;**19**:161-165

[64] Kahneman D. Maps of bounded rationality: A perspective on intuitive judgment and choice. In: Frangsmyr T, editor. Les Prix Nobel: The Nobel Prizes, Stockholm. Sweden: Nobel Found; 2002. pp. 449-489

[65] Gigerenzer G, Gaissmaier W. Heuristic decision making. Annual Review of Psychology. 2011;**62**:451-482

[66] Raab M, Gigerenzer G. The power of simplicity: A fast-and-frugal heuristics approach to performance science. Frontiers in Psychology. 2015;**6**:1672

[67] Kripke SA. Semantic analysis of modal logic (abstract). The Journal of Symbolic Logic. 1959;**24**:323-324

Chapter 9

Methodological Approach in the Development of Specific Games in Elite Soccer

Javier Vilamitjana, Julio Calleja-Gonzalez and Diego Marqués-Jiménez

Abstract

In games such as soccer, where the stability and the possibility of replicating game situations are complex, teams and players continually deal with a highly unstable cooperative and non-cooperative environment. Thus, synchronized cooperation among players during training sessions is a fundamental factor, which many times contributes to a team's success. In this context, there are some specific drills that attempt to challenge and create meaningful contexts in order to simulate match situations as closely as possible. Small-sided games are play-sport situations in which all elements of the game interact together in a flexible manner. However, there are a variety of small-sided games in elite soccer, such as possession games and positional games, which may present specific characteristics and stimulate different physical-physiological demands. An adequate selection and implementation of these games may help coaches to promote positive adaptations and performance improvements. Thus, this chapter provides practical tips to modulate the physical-physiological responses and technical-tactical requirements of the players using a variety of game formats during soccer training sessions.

Keywords: soccer, ball possession, small-side games, possession games, positional games

1. Introduction

Game theory plays a key role in the applied social sciences. Concretely, it has been used in consequence analysis of "decision-making" in the tactical performance of any given individual in team sports, which is a fact to be considered for the players of the same team as for their opponents [1]. In particular, games such as soccer where the stability and the possibility of replicating game situations are complex, teams and players continually deal with a highly unstable cooperative and non-cooperative environment [2]. From this perspective, players and teams are conceptualized as dynamic, intricate systems, interacting in a nonlinear fashion with the environment. Therefore, exposure to challenging and meaningful contexts pushes the exploration and discovery of new synergies, promoting co-adaptive processes between players and transforms sports into dynamic entities [2]. Thus, synchronized cooperation among players is a fundamental factor, which many times contributes to a team's success. This phenomenon will be substantially more favorable than the fictitious sum of the technical and motor qualities inherent to each player belonging to that team [1]. In this context, there are some specific drills such as "small-sided soccer games" that attempt to challenge and create meaningful contexts in order to simulate match situations as closely as possible.

Small-sided games (SSGs), also referred to in the literature as "game-based training" [3], are play-sport situations [4] in which all elements of the game interact together in a flexible manner [5]. In addition, there are a variety of specific game approaches in elite soccer. The use of such games in professional environments is based on the premise that greater performance improvements are achieved when the specific demands of the sport are transferred [6, 7]. Several studies have shown that the physiological responses of different games can be modified by manipulating variables such as number of players per team [8, 9], modification of certain rules [9], relative area per player [6, 10, 11], comparison with competition [12–14], floaters [15], among others. The authors concluded that these games enable players to get as close as possible to real competitive situations. On account of this, the physical, physiological, technical, and tactical demands of a match can be reproduced to a greater extent.

Moreover, using a weekly pattern with different formats of SSGs in each training session may modulate the player's training load during the micro-cycle [16], which allows a short tapering strategy to face the match with enough energy. Therefore, it might be utilized as a strategy for maintaining or optimizing players' physical performance during the season [17].

As a consequence, coaches and performance staff have made emphasis on and proposed an infinite number of exercises, all of which count with variations in pitch size, number of players in each team, game instructions, and different designs in the shape of the pitch to be used [12, 14]. However, to the best of the author's knowledge, no previous references have been published in order to classify and contrast each kind of game used in elite soccer to train players to reach the physical, technical, and tactical demands of the match. In this context, the present chapter attempts to classify and categorize the different types of specific games and describe their characteristics and the physical-physiological demands.

2. The "Conventional Small-Sided Games"

The SSGs represent one of the most common training elements in soccer at any level and age, as they allow the simultaneous development of technical-tactical contents together with physical goals [18]. In particular, the player's responses during the performance of SSGs have been extensively studied by different authors [12, 19, 20]. In this sense, the advantage of carrying out these games is to replicate real competition situations as closely as possible and thus be able to reproduce very similar physical, physiological, technical, and tactical demands of the game [6, 13]. These types of games can be configured according to a variety of components such as number of players, space orientation, individual interaction space, and balance (whether the teams have the same number of players or are unbalanced by floaters) [12].

The SSGs are collaboration-opposition games [4], which possess a space orientation (one team defending and the other attacking opposite goals), that count with a *"sequence of attacking"* (possession) and *"defending"* (out-of-possession). In this way, two forms of transition are originated: 1) from possession to non-possession of the ball, or 2) from non-possession to possession of the ball [21].

2.1 "Space orientation" and "Ball Possession Sequence"

The orientation of the space in SSGs is defined as the presence or absence of space targets in which a particular aspect that characterizes these types of games is defined [4]. On the one hand, when there are no goals or no scoring zones and the aim is to keep ball possession, it is called *"non-oriented area"* [12]. On the other hand, when there are goals and/or marking zones, but each team attacks and defends them, it is a game with an *"oriented area"* [12]. Likewise, when goals or scoring zones are located in the field, each team knows which one they are defending and which they are attacking. This aspect aligns the game in a way called *"polarized area"* and forges preferential *"paths of action"* [12, 22]. SSG formats with a non-oriented area encourage players to cover longer distances at higher running speeds, whereas SSGs with a polarized area increase the time in which the ball is out of play [23]. In consonance with these fundamentals, Casamichana et al. [24] obtained a greater cardiac response in non-oriented space tasks when 4 vs. 4 formats (100 m^2 per player) with a modification in orientation were compared (**Figure 1**).

2.2 "Relative Playing Area per Player" and "Number of Players": Physiological responses

Relative playing area per player should be calculated by dividing the total play surface of each SSG among all players (m^2 per player) [25]. Some authors have found that SSGs' intensity can be manipulated by modifying the relative playing area per player and the player numbers. For instance, Hill-Hass et al. [26] examined the effect of three formats of player numbers (2 vs. 2, 4 vs. 4, and 6 vs. 6) with the same relative playing area (150 m^2 per player) on physiological patterns and rate of perceived exertion (RPE). As a relative pitch area per player decreased, the overall physiological performance and RPE increased. In fact, for a fixed pitch area, the lower number of players, the higher the RPE was [19, 23, 26]. As a counterpart, Rampinini et al. [27] performed research with a variety of relative playing areas per player, although the number of players was always equal. Thus, it was clearly shown that in the 3 vs. 3 and 6 vs. 6 formats, increase in pitch size led to higher physiological parameters and perceived intensities (heart rate, blood lactate concentration, and RPE).

Other researchers compared conventional formats (4 vs. 4, 6 vs. 6, 8 vs. 8, from 71 to 106 m^2 per player) with 10 vs. 10 small game (311 m^2 per player) and competitive matches (1-4-3-3 formation), concluding that only the 10 vs. 10 format allowed players to

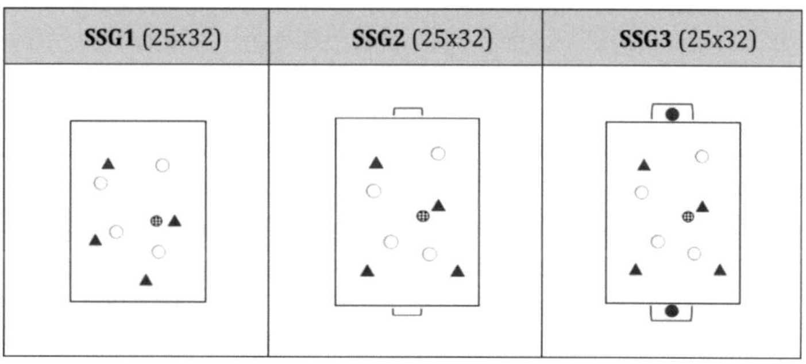

Figure 1.
SSG formats of 4 vs. 4. SSG1: Non-oriented space. SSG2: Oriented space without goalkeepers and with small goals. SSG3: Oriented space with goalkeepers and official goals (extracted with permission from Casamichana & Castellano [10]). In brackets, the width and length of the pitch used in each design (in meters).

reach similar intensities and distances to those obtained during matches, whereas the 4 vs. 4 format exhibited the greatest difference in mechanical work and the least difference on distance above 14.4 km/h [28]. Previous studies conclude that increasing the number of players (and concomitantly relative playing area per player) increases total and high-speed distance (>14.4 km/h) during SSGs [29–31]. Owen et al. [32] also reported that these formats do not induce high-speed movements compared with the ones with larger relative playing area per player and therefore more players (9 vs. 9 to 11 vs. 11).

2.3 Floater players during small-sided games

The *"floater"* is a special player who belongs to the team in possession of the ball during the development of the SSGs, allowing teams to obtain a numerical superiority [33]. The load imposed on regular players when performing SSGs with different numbers and distributions of floaters has been studied in other studies [34, 35]. Sánchez-Sánchez et al. [34] observed that the introduction of interior and exterior floaters reduces the RPE, the heart rate response, and the number of dribbles with respect to the control situation, without the presence of floating players. Regarding floater's performance, Lozano et al. [35] compared 4 vs. 4 + 2 format with 8 vs. 8 + 1 (area: 44–75 m^2 per player) and official matches, reporting that total distance, high-intensity distance (>14.4 km/h), sprint distance (>21 km/h), accelerations (>2 m/seg^2), and decelerations (< −2 m/seg^2) were lower in the floaters compared with regular players. In this way, Rábano-Muñoz et al. [36] showed that floaters registered lower external training loads in comparison to regular players with respect to peak velocity and maximum heart rate.

2.4 Technical-tactical outcomes

Regarding rules inherent to tactical or strategic outcomes, there are studies with special considerations. For example, Fradua et al. [37] extrapolated SSGs' sizes from the actual pitch (11 vs. 11) to investigate parameters related to tactics in the game and concluded that pitch size is a variable that influences ball possession. Thus, the variation in pitch size can create favorable (and unfavorable) conditions for attack and defense [38, 39]. Other authors concluded that SSGs with 9 vs. 9 to 11 vs. 11 (218–336 m^2 per player) are also more suitable to simulate most of the specific technical profile (passing actions, such as long-distance and penetrative passes), while 4 vs. 4 format (94 m^2 per player) makes more emphasis on more short distance passes during the activity [32]. In addition, Casamichana & Castellano [10] examined physical, physiological, and motor responses and RPE during different SSGs (~75–275 m^2) while the number of players per team was kept constant: 5 vs. 5 plus goalkeepers (the participants were 10 male youth soccer players). When the individual playing area was larger, the effective playing time and the physical-physiological patterns were higher, while certain motor behaviors were observed less frequently (interception, control and dribble, control and shoot, clearance, and putting the ball in play). The authors concluded that the size of the pitch should be taken into consideration when planning training drills, as it influences the intensity of the task and the motor response of players.

Furthermore, different functional movement behaviors emerged as a consequence of the manipulation of the environmental situation. For instance, Gonçalves et al. [40] compared the players' positioning dynamics, manipulating the number of opponents and teammates (numerical inequality) during professional and amateur SSGs. The participants played 4 vs. 3, 4 vs. 5 and 4 vs. 7 games (109–171 m^2), where one team was confronted with low-superiority, low- and high-inferiority situations, and their opponents with low, medium, and high-cooperation situations. The conclusions revealed that the increasing number of opponents was effective to overemphasize the need to use local information when deciding a position making process in professional players (they presented higher regularity in movement behavior as the number of opponents increased). Conversely, amateur players still rely on external informational feedback (when cooperation was increased, more spatial organization was obtained and players' local perceptions were emphasized).

In this context, *"conventional SSGs"* could facilitate the development of a core tactical concept with an appropriate game context, although this will depend on its design [19]. For our methodology proposal, this new line of analysis leads to another approach in specific soccer games: the *"Possession Games."*

3. The possession games

Not only are the SSGs' traditional approaches commonly used in the soccer world [9, 20], but other games are also practiced with the main goal of progressing with the possession of the ball by the attacking team. Among them, we can find possession games (POGs), which are relatively similar to *"conventional SSGs,"* yet nonetheless, have a number of different characteristics. During SSGs, the aim of the task is to maintain ball possession, but the disposition of the players is not preset and the occupation of the spaces is not predetermined, while in POGs, the same spaces are intelligently covered [14]. In the latter, the players who maintain possession of the ball are positioned in such a way that the interrelation among them and the space is as efficient as possible, stimulating the development of individual and collective concepts for the understanding of the game (**Figure 2**).

Effectively, the fundamental objective of this type of exercise is to generate free spaces by means of individual and collective movements, which allow the game to

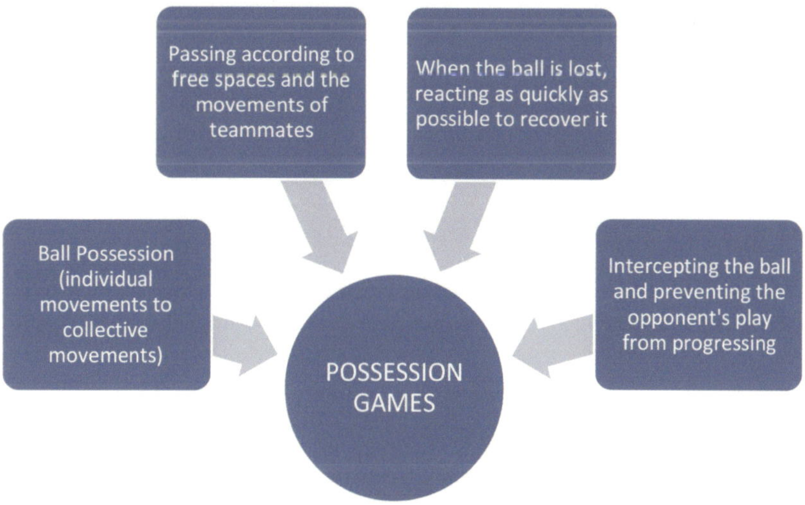

Figure 2.
Basic principles of the possession games: a) ball possession by means of individual movements toward collective movements (e.g., deep movements, diagonal movements, etc.), b) passing, depending on the free space and the different movements (e.g., lateral pass looking for width, vertical pass looking to progress, deep pass between lines, etc.), c) ball recovery, which aims to go to the opposing player in possession of the ball, and finally, d) intercepting and thus preventing the opponent from progressing in the attack, Vilamitjana et al. [41].

"progress" with greater fluidity in a particular direction (**Figure 2**). This means that the creation of movements to generate free spaces will be useful to make passes toward the mentioned spaces in order to generate collective movements. In relation to this concept, in his book *"Me gusta el Fútbol"* [41], Johan Cruyff synthesizes the game of soccer with this following phrase: *"Don't run too much since soccer is played with the brain. You have to be in the right place at the right time, not before or after."* This principle gives rise to factors inherent to strategy and tactics, which have a greater transfer toward specific match situations [14]. In this line, the practice of POGs will be most effective in pitches designed with different shapes and spaces (**Figure 3**).

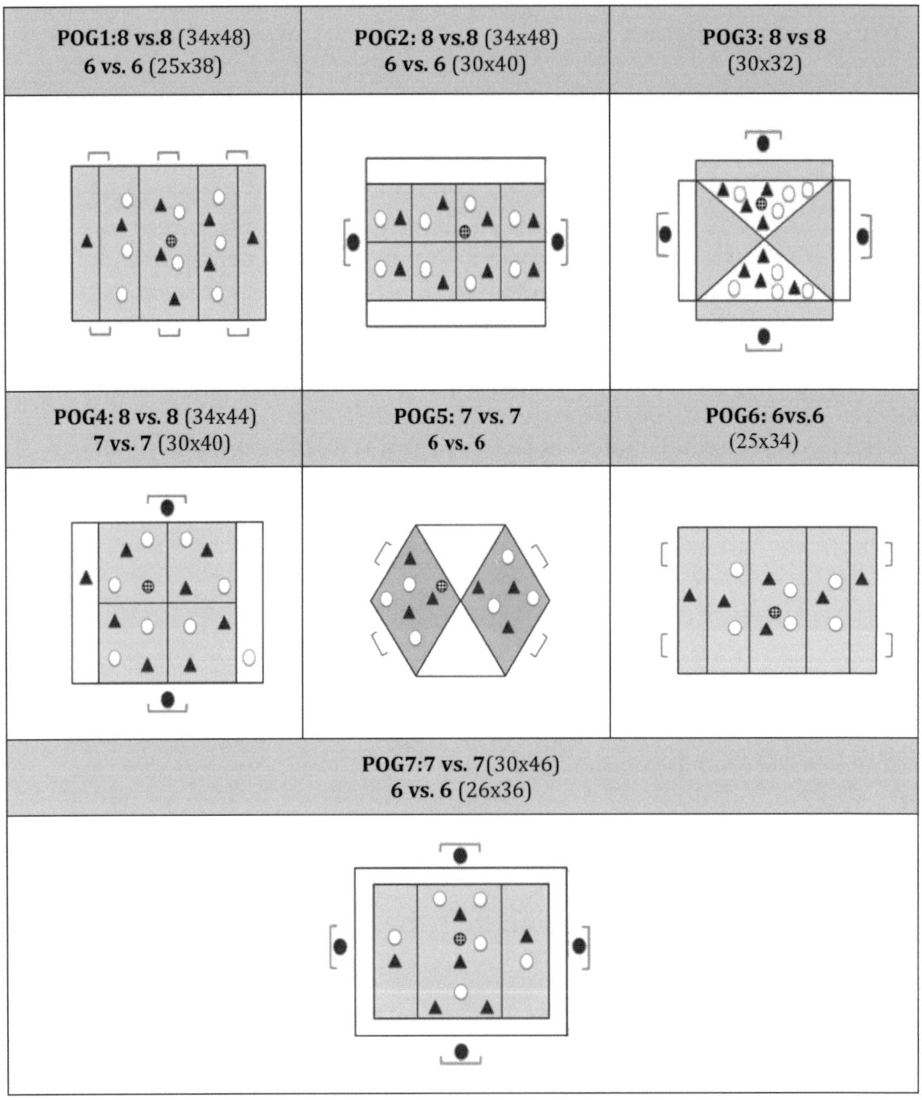

Figure 3.
POG designs for the three formats (6 vs. 6, 7 vs. 7 and 8 vs. 8) studied by Vilamitjana et al. [14]. In brackets, the width and length of the pitch used in each design (in meters).

3.1 Possession games´ comparison with small-sided games and competition: Physiological responses

There are studies using POGs with different designs and player numbers in comparison with official games. For instance, Vilamitjana et al. [14] compared three POGs formats (65–110 m^2 per player) with official matches (1-4-2-1-3 and 1-3-4-3), and then differentiated the final performance according to the positions occupied by the players on the field (**Figure 3**). They concluded that the cardiovascular response in 6 vs. 6 and 7 vs. 7 was match-compatible and related to cardiovascular performance, the mean values in POGs were no different from the matches, with the exception of 8 vs. 8 formats. In relation to the high-intensity running and sprinting work rate, the authors found low percentages in both patterns when compared with competition.

In Ref. [31], the authors compared SSGs with POGs in three different formats. They concluded that both formats with a smaller number of players (5 vs. 5, 7 vs. 7; 73 and 98 m^2 per player, respectively) do not induce high-speed movement compared with the ones with larger pitches and more players (10 vs. 10; 135 m^2 per player). This effect was due to a larger pitch area and less pressure received from opponents, with a greater number of options for passing the ball among players. The main task during both games was that players could not progress with more than two touches (no difference in the designs for each format were detailed, as the only difference described was that POGs were played in a "non-oriented area" and the SSGs were played with goalkeepers and goalposts). The results determined that very high-intensity distance (19.8–25.2 km/h) covered was higher in SSGs in relation to POGs (no significant difference was found at >14.4 km/h), with a larger number of high-intensity accelerations (> 3.0 m/sec^2) and decelerations (< −3.0 m/sec^2) in favor of SSGs when compared with POGs for the format 5 vs. 5.

As a counterpart of this, a descriptive study was made with typical POGs training sessions (oriented area games) and conventional SSGs designs (**Figure 4**), 5 vs. 5

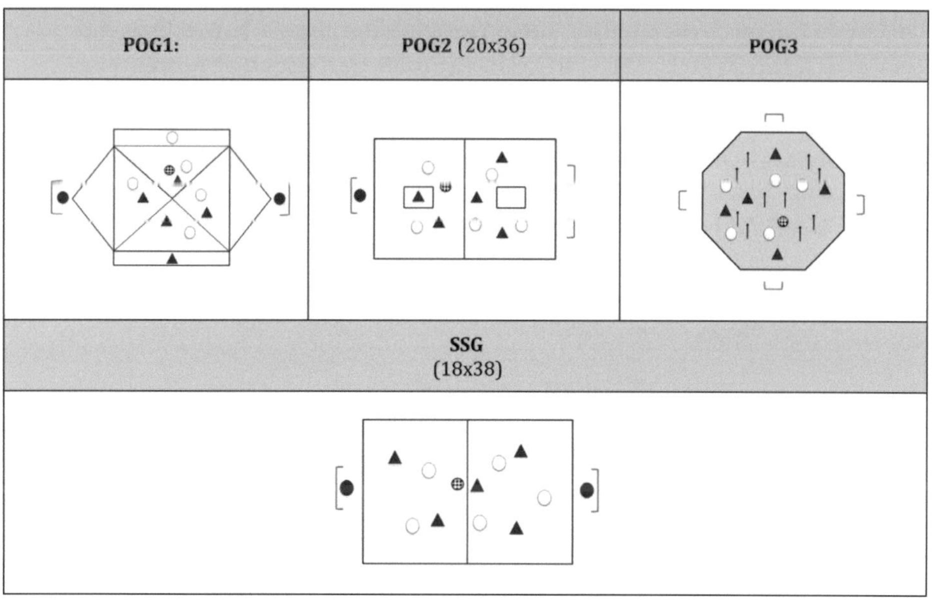

Figure 4.
Possession games and conventional small-sided game designs and diagram representation for 5 vs. 5 formats studied by Vilamitjana et al. [42]. In brackets, the width and length of the pitch used in each design (in meters).

formats (70 m^2), in comparison with official matches (1-3-4-3 and 1-4-2-1-3) [42]. The analysis of the data described higher performance in POGs during seven of the forehead nine study variables (total distance, player-load, high-intensity work rate, high-speed intensity work rate, number of runs in high-intensity running, and high-speed running and maximal speed) except in high-intensity accelerations (> 3.5 m/seg^2) where the SSGs values were higher than POGs, while in high-intensity decelerations (< −3.5 m/seg^2), no significant differences were obtained. When the data were discriminated by field position, central defenders and midfielders obtained similar values to competition situation in the variables of high-intensity work rate (> 14.9 km/h) and very-high-intensity work rates (> 19.9 km/h). The current findings suggest that POGs are a very interesting tool to stimulate the physical demands to which players will be exposed to during matches. Moreover, SSGs could be utilized as an exercise with greater intentionality when it involves stimulating the accelerations that the player performs during a specific execution time [42].

3.2 Floating players during possession games

There are not many studies performing POGs with different numbers of floaters. Asian-Clemente et al. [43] compared two formats (non-oriented area), with two floaters (both exercises were designed with the same relative area per player; 81 m^2 per player) and official matches (**Figure 5**). In both formats, players were divided into three teams. The POGs were classified on whether they exhibited a change of play area (POGca) or if there was no change of play area (POGnc). In both cases, floaters always have an offensive role, playing with teams in possession of the ball. During POGca, two teams played in a certain area (5 vs. 5 + 2) and when the attacking team scored 7 passes or the defensive team recovered the ball, they had to perform a pass to another zone where the third team was waiting the pressure of one of them. In POGnc, two teams played against one team (10 vs. 5 + 2). The aim of both games was to maintain ball possession until another team intercepted the ball or kicked it outside the pitch. Next, they had to exchange roles with the other team of five players (**Figure 5**). The authors concluded that POGca performed higher values regarding a greater total distance, high-speed, peak speed, and number of accelerations (> 3 m/seg^2) and decelerations (< −3 m/seg^2) than POGnc. Comparing both exercises with match situation, POGs showed a significantly higher speed and an increased number of accelerations-decelerations [43].

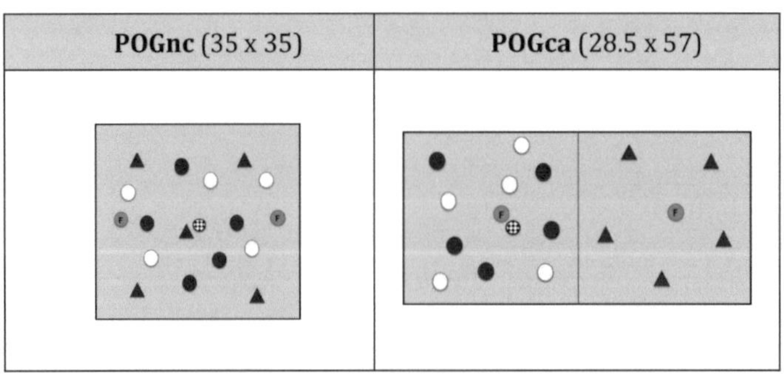

Figure 5.
Possession games designs with 5 vs 5 + 5 plus 2 floaters (extracted with permission from Asian-Clemente et al., 2021). In brackets, the width and length of the pitch used in each design (in meters).

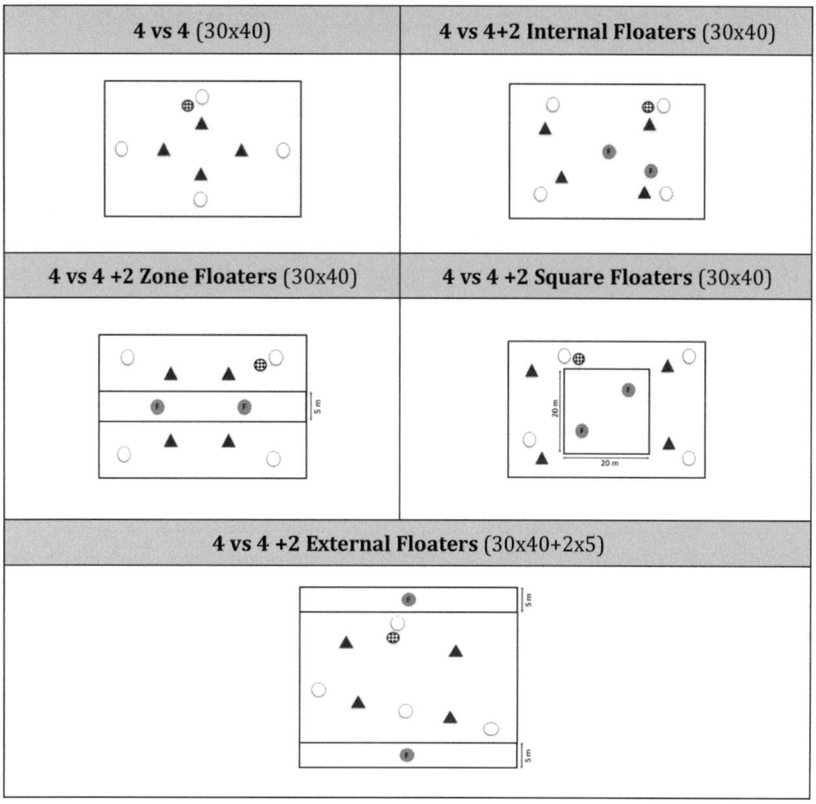

Figure 6.
Possession game designs with 4 vs 4 and 4 vs 4 plus 2 floaters formats (extracted with permission from Asian Clemente et al., 2022). In brackets, the width and length of the pitch used in each design (in meters).

In another research, Lacome et al. [33] performed a study comparing POGs vs. SSGs with one floater (61–120 m^2 per player). The authors reported that total distance, high-intensity distance (> 14.4 km/h), accelerations (>2 m/seg^2), decelerations (<−2 m/seg^2), and changes of direction were lower in the floaters compared with regular players independently of SSGs or POGs designs.

Another relevant information was recently obtained by the same authors utilizing 4 vs. 4 format (non-oriented area) with the incorporation of two floaters (∼120–150 m^2 per player), who always assumed an offensive role during ball possession tasks [44]. They demonstrated that regular players completed a greater total distance and distance covered between 14 and 17.9 km/h than without floaters (**Figure 6**).

Regarding technical profiles during these exercises, studies from Mallo and Navarro [45] found that the introduction of wildcards in 3 vs. 3 formats (where the objective was to maintain possession of the ball) significantly reduced the number of contacts with the ball. Additionally, they found that error percentage in passes performed by the players retained no modification in heart rate response or locomotor activity.

3.3 Technical-tactical outcomes

Concerning inherent factors related to tactical and technical skills, there are some authors who studied the specific actions associated with each match effort during *"ball possession"* and *"out-of-possession"* [46]. As previously described, the POGs' principles give rise to factors as *"concepts"* related to strategy and tactics, which have a greater transfer capacity toward specific match situations (**Figure 2**). The different variety of shapes and spaces facilitate the implementation of specific movement patterns. For instance, in *"the hexagon shape"* (**Figure 4**, POG#1), the attackers carry out actions of overlapping (player runs from behind to in front of or parallel to the player on the ball) to progress the collective movements and perform depth passes to opposite box to break into the adversary team lines. Meanwhile, defenders cover the spaces, closing down the opponent players, trying to cut out passes from them. In other designs of POG, *"the double diamond shape"* (**Figure 3**, POG#5) contemplates the attackers' actions of movements with depth (diagonal and vertical movements), swiftly enabling different sides profiles, with visual optimization and technical skill. Finally, the players must find a suitable pass to the opposite side considering always viewing the space beyond their immediate area. Moreover, the defensive team has specifics tasks, as a main one recovery of the ball by running in a collective way toward opponent players, attempting to intercept the ball and change spaces immediately to another triangle or diamond; in this manner, a possession sequence begins. There is a relevant concept to be considered when the team loses the ball: the team has no other choice but to reorganize and jump in to put pressure into recovering the ball as quickly as possible (**Figure 2**).

A new tactical dimension starts when players take up specific role positions, trying to have a gravitational effect on their opponent by superiority, when a new approach of specific games is proposed: the "Positional Games."

4. The positional games

The positional games (PGs) are performed with the objective of team ball possession in which the players have priority action areas based on their position in competition, where playing space is adapted to the player's usual context in matches, but without restricting the players' spatial exploration during the tasks [15]. These positional games require selected roles to position themselves intelligently (this design usually uses vertical and horizontal lines on the pitch, with each player assigned to a zone), and the team works dynamically and collectively in accordance (**Figure 7**).

Ball possession takes on a more tactical sense in the PGs: they attract the opponent to press in such a way that they must press on the offensive (persuading action), demonstrating at some point certain vulnerability on the defensive side. This will be the moment to act speedily to confront the opponent's moves and thus, finally, break the defense originated by opposing team (**Figure 7**). Therefore, ball possession is a constructed phenomenon, because it is a possession that aims to destabilize the opponent, eliminate rivals, and condition their defensive balance, forcing them to adjust constantly to these elements and thus play at their mercy rather than play as the rival would wish to do so.

In general, PGs are utilized with *"floating players,"* who encourage ball retention and generate numerical superiority for the team during ball possession [15]. The floaters intervene only on the offensive side, placing themselves in intelligent positions (most appropriate positioning for tactical resolution), thus favoring ball possession and attacking progression (**Figure 7**) [15].

Head coach and former player Gabriel Heinze considers PGs as *"A style of play, a team identity, a way of perceive training and competition, all of which require conviction on*

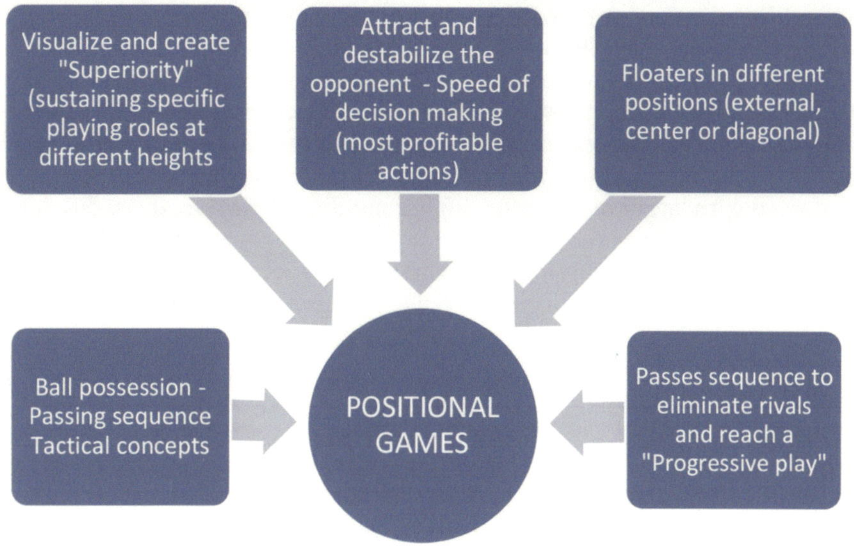

Figure 7.
Basic principles of the positional games. a) Ball possession by means of individual movements toward collective movements, b) visualize the playing context, retaining the ball and generating superiority in small spaces, c) attract the opponent and persuade them to press, d) the most appropriate floaters positioning for tactical resolutions during the attack, and e) passing sequence with other teammates or, alternatively, the possibility of ball conduction and progression in the game (Vilamitjana, J. & Heinze, G.).

the part of the coach." At the same time, they also require faith and trust from the players; otherwise, these types of games will be difficult to implement.

4.1 Relative playing area per player and player numbers: Physiological responses

To the best of the author's knowledge, no previous references have been published in order to contrast PGs data with the other games. In one practical experience, a comparison of three formats of PGs (68–81.6 m^2 per player, polarized area, 1–2 floaters) with official matches was carried out (**Figure 8**). On the one hand, the findings revealed that some metrics decreased progressively from PG1 to PG3 (PG1 > PG2 > PG3): 9.2–4.9% in meters per minute, 8.8–7.7% in player-load, and 4–3.2% in mean heart rate (171.8–164.8–159.6 beats per min), respectively. On the other hand, high-intensity patterns increased progressively from PG1 to PG3 (PG3 > PG2 > PG1): distance above 19.9 km/h (4.7–9.2%), and maximal speed (3.2–5.6%).

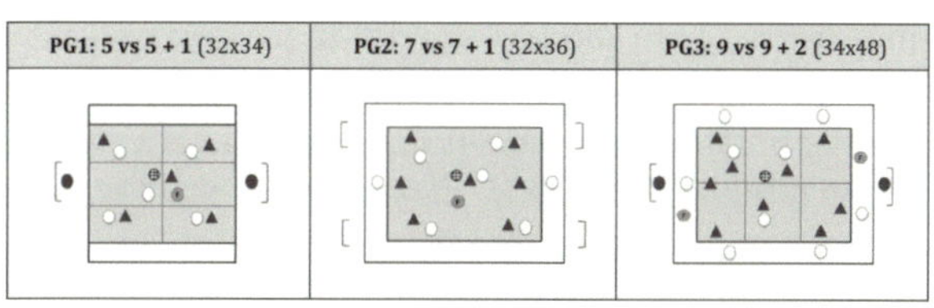

Figure 8.
Positional game designs with floaters format studied by Vilamitjana, J., & Heinze, G. (under revision). In brackets, the width and length of the pitch used in each design (in meters).

When PGs were compared with match situation, obtaining lower values of work rate profiles for each format on which the study was undertaken. Only parameters such as sprints, accelerations, and decelerations were higher in all formats compared with official matches. Finally, the conclusion was similar to those determined with SSGs: increasing the number of players and relative playing area per player would induce high-speed patterns, and it seems that when smaller games are compared with larger ones, with a higher number of players, these did not reach similar intensities and distances to those obtained during matches. In this context, it should be made clear that physical performance is important, but the tactical-cognitive conception that the players carry is what prevails the most in this type of game.

4.2 Technical-tactical outcomes

Beyond the physical and physiological performance, in order to achieve the principles described above, PGs have a high level of cognitive and technical skill requirements (**Figure** 7). To begin with, every player has a direct (with the ball) or indirect (without the ball) responsibility in relation to a defined tactical concept for each playing position. On the one hand, players without the ball must occupy certain spaces to provoke a determined, sought-after behavior in the opponents (attract the opponent and persuade them to press), either by jumping in to pressure the ball carrier or by maintaining proximity to their teammate with their mark. This facilitates the passing sequence with other teammates or, alternatively, the possibility of ball conduction and reach a *"progressive play."* On the other hand, the player with the ball has the intention of attracting the opponent's pressure to find free players located at different heights of the field (**Figure** 7).

Another relevant concept to be considered in PGs is "superiority" (numerical, qualitative, and positional) [47]. Numerical superiority is a team with possession overload in any area of the pitch (floaters help the team to generate this aspect). Qualitative superiority is when a player who is superior to their direct opponent isolates them in a 1 vs. 1 or 2 vs. 2 situations (it is relevant the movements from players without ball). Positional superiority involves getting players into positions between or behind the opposition lines, where they are most likely to have time and space relative to the ball. Consequently, the aforementioned superiority is more likely to affect the game (trying to find the free man directly or indirectly). Any player in a team using positional play can achieve one of these types of superiority, but everyone must sustain his or her specific playing role during the game (**Figure** 7). It is essential that this tactical concept is built from the back (first tactical line). For this reason, a fundamental principle of its idea of play is that the ball comes out cleanly from the defenders: From the first line, the different game positions will try to retain the ball generating superiority in small spaces (progress the ball forward through the creation of triangles or diamonds that give the ball-carrier space and several passing options at any given time). For instance, center-backs moving wide, trying to provoke that the forward of the opposing team jump in to press, which in turn creates a passing lane into the midfielders (in a higher position). Players need to be ready to move based on the movement of a teammate. This creates constant rotations that aim to disrupt the opposition.

The *"out-of-possession"* is a very important phase inside PGs, because on account that the team has to reorganize and jump in rapidly to put pressure on the opponent (in coordination with all lines of tactical positioning). It is considered a similar

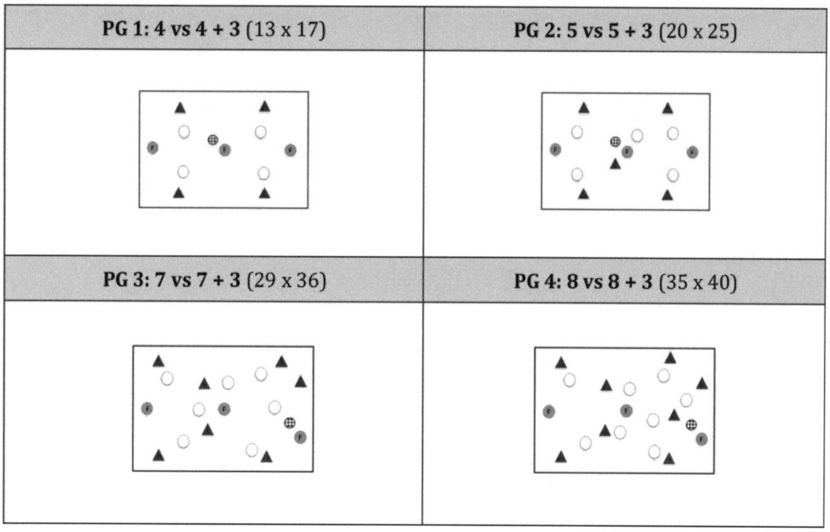

Figure 9.
Positional game designs with floaters format (extracted with permission from Casamichana et al., 2018). In brackets, the width and length of the pitch used in each design (in meters).

concept to the one described previously in POGs, even though there exist specific out-of-possession strategies employed by the teams using this style of game.

4.3 Floater players during positional games

The implementation of floaters during PGs facilitates ball possession and consequently, generates numerical superiority for the team retaining the ball [48]. Casamichana et al. [15] studied the kinematic demands imposed on floaters and regular players in addition to comparing the demand imposed on wildcards in different PGs formats (~20–74 m^2 per player respectively) (**Figure 9**). The main conclusion that resulted was that floaters imposed lower intensities than regular players in high metabolic distance (> 25.5 W.kg), but this difference was smaller in PG1 and larger in PG4 format. Moreover, the demand imposed on floater players in PGs that were studied also revealed the following differences: there were a greater number of accelerations and decelerations in the smaller format (PG1) compared with the larger formats (PG3-PG4), while total distance covered and high metabolic distance were greater in the larger formats (PG3-PG4) compared with the smaller ones (PG1-PG2) (**Figure 9**).

5. Conclusions

There are a considerable number of designs within the three types of games described in this chapter to be taken advantage of. During these specific types of exercise, "*ball possession sequence*" is a typical common denominator to be considered. In particular, SSGs appear to be a basic concept in which all the game's elements interact in a flexible way: the aim of the task is to maintain ball possession, but the disposition of the players is not preset, and the occupation of the spaces is not predetermined. In another approach, there are POGs where the players who maintain possession of the ball are positioned in such a way that the interrelation among them and the space is as efficient as possible. Hence, free spaces are generated by individual and collective movements, which make the ball possession *"progress"* with greater fluidity, with a particular direction and purpose. Finally, we have PGs that count with a higher level of cognitive and technical skill requirements. In these games, ball possession takes on a more tactical sense, in which the players have priority action areas based on their position in competition.

5.1 Practical applications

- This chapter reveals that coaching staff may modulate the physical-physiological responses and technical-tactical requirements of the players, using a variety of game formats during soccer training sessions.

- The manipulation of different variables in each game (number of players, balance, relative playing area per player, space orientation, among others) may influence the players´ responses.

- The "conventional SSGs" could be an exercise, which could be practiced with greater intentionality, especially when desired result is to stimulate the number of accelerations. The POGs are actually, for all intents and purposes, an exceptional approach to stimulate most of the physical demands that players are subjected to during competition, and at the same time, PGs apply tactical concepts with the intervention of different physical patterns.

- In this context, the manner in which soccer coaches and physical trainers prescribe training definitely has a key role in helping players' development for successful performance. As a consequence, an adequate selection and implementation of these games may help coaches to promote positive adaptations and performance improvements.

5.2 Future research lines

Factors inherent to certain conditioning components such as the associations that have the origins during POGs together with their transfer to match situations should be further investigated in an empirically way. Likewise, PGs have been designed for the development of tactical concepts, but more scientific data on the physiological response to this physical load and technical requirements are still needed. Moreover, the cited studies in this chapter were mainly performed with professional or elite soccer players, and as a result, future research is warranted to likewise understand how youth players are coping with different games.

Author details

Javier Vilamitjana[1*], Julio Calleja-Gonzalez[2] and Diego Marqués-Jiménez[3]

1 Soccer Research Group, Friends Club – CDA, Buenos Aires, Argentina

2 Department of Physical Education and Sports, Faculty of Education and Sport, University of the Basque Country, UPV/EHU, Vitoria-Gasteiz, Spain

3 Faculty of Education, University of Valladolid, Soria, Spain

*Address all correspondence to: vilamitjana@yahoo.com

References

[1] Sindik JI, Vidak N. Application of game theory in describing efficacy of decision making in Sportsman's tactical performance in team sports. Interdisciplinary Description of Complex Systems. 2008;**6**(1):53-66

[2] Pol R, Balagué N, Ric A, et al. Training or synergizing? complex systems principles change the understanding of sport processes. Sports Medicine. 2020;**6**:28

[3] Gabbett T, Jenkins D, Abernethy B. Game-based training for improving skill and physical fitness in team sport athletes. International Journal of Sports Science and Coaching. 2009;**4**(2):273-283

[4] Parlebas P. Juegos, deporte y sociedad. Léxico de praxiología motriz. Barcelona: Paidotribo; 2001

[5] Wein H. Fútbol a la medida del niño. Vol. 1. Madrid: Gymnos; 1995

[6] Dellal A, Chamari K, Owen A, Wong D, Lago-Penas C, Hill-Haas S. Influence of the technical instructions on the physiological and physical demands within small-sided soccer games. European Journal of Sport Science. 2011; **11**:353-359

[7] Little T. Optimizing the use of soccer drills for physiological development. Strength and Conditioning Journal. 2009;**31**(3):67-74

[8] Brandes M, Heitmann A, Müller L. Physical responses of different small-sided game formats in elite youth soccer players. Journal of Strength and Conditioning Research. 2012;**26**(5):1353-1360

[9] Hill-Haas S, Dawson B, Coutts A, Rowsell G. Physiological responses and time-motion characteristics of various small-sided soccer games in youth players. Journal of Sports Sciences. 2009; **27**(1):1-8

[10] Casamichana D, Castellano J. Time–motion, heart rate, perceptual and motor behaviour demands in small-sides soccer games: Effects of pitch size. Journal of Sports Sciences. 2010;**28**(14):1615-1623

[11] Porres D, Paz Fernández J, Fernandez Gonzalo R, Cervera J, Yagüe CJ. Variabilidad de la carga fisiológica en los pequeños juegos de fútbol en función del espacio. Apunts Educación Física y Deportes. 2010;**102**:70-77

[12] Casamichana D, San Roman J, Calleja J, Castellano J. Cuantificación de las cargas durante los juegos reducidos. Los juegos reducidos en el entrenamiento del fútbol. 1st ed. España, Barcelona: Fútbol de Libro; 2015

[13] Dellal A, Owen A, Wong D, Krustrup P, Van Exsel M, Mallo J. Technical and physical demands of small vs. large-sided games in relation to playing position in elite soccer. Human Movement Science. 2012;**31**:957-969

[14] Vilamitjana J, Heinze G, Verde P, Calleja-González J. Comparison of physical performance between possession games and matches in professional football. Apunts. Educación Física y Deportes. 2020;**141**:75-86

[15] Casamichana Gómez D, Gómez Díaz A, Cos Morera F, Martín GA. Wildcard players during positional games. Educación Física y Deportes. 2018;**133**:85-97

[16] Marqués-Jiménez D, Sampaio J, Calleja-González J, Echeazarra I. How different are soccer training sessions

based on small-sided games? A cluster analysis to explore perceived exertion and training load. Acta Gymnica. 2022; **52**:e2022.005

[17] Guridi Lopategui I, Castellano Paulis J, Echeazarra EI. Physical demands and internal response in football sessions according to tactical periodization. International Journal of Sports Physiology and Performance. 2021;**16**(6):858-864

[18] Halouani J, Chtourou H, Gabbett T, Chaouachi A, Chamari K. Small-sided games in team sports training: A brief review. Journal of Strength and Conditioning Research. 2012;**28**(12): 3594-3618

[19] Hill-Haas S, Dawson B, Impellizzeri F. Coutts, A. physiology of small-sided games training in football: A systematic review. Sports Medicine. 2011;**41**(3):199-200

[20] Casamichana D, Castellano J, Calleja-González J, San Román J, Castagna J. Relationship between indicators of training load in soccer players. Journal of Strength and Conditioning Research. 2013;**27**:369-374

[21] Orta A, Pino J, Moreno I. Propuesta de un método de entrenamiento universal para deportes de equipo basándose en el análisis observacional de la competición. Lecturas: EF y Deportes; 2000

[22] Castellano J. Confección de tareas de entrenamiento en fútbol. Español: El Entrenador; 2005

[23] Köklü Y, Sert Ö, Alemdaroğlu U, Arslan Y. Comparison of the physiological responses and time-motion characteristics of young soccer players in small-sided games: The effect of goalkeeper. Journal of Strength and Conditioning Research. 2015;**29**(4): 964-969

[24] Casamichana D, Castellano J, González-Morán A, García-Cueto H, García-López J. Demanda fisiológica en juegos reducidos de fútbol con diferente orientación del espacio. Revista Internacional de Ciencias del Deporte. 2011;**23**(7):141-154

[25] Parlebas P. Elements of Sociology in Sport. Málaga: Unisport Andalucía; 1988

[26] Hill-Haas S, Rowsell G, Dawson B, Coutts A. Acute physiological responses and time-motion characteristics of two small-sided training regimes in youth soccer players. Journal of Strength and Conditioning Research. 2009;**23**(1): 111-115

[27] Rampinini E, Impellizzeri F, Castagna C, Abt G, Chamari K, Sassi A, et al. Factors influencing physiological responses to small-sided soccer games. Journal of Sports Sciences. 2007;**25**: 659-666

[28] Lacome M, Simpson B, Cholley Y, Lambert P, Buchheit M. Small-sided games in elite soccer: Does one size fits all? International Journal of Sports Medicine. 2017;**17**:1-24

[29] Castellano J, Casamichana D, Dellal A. Influence of game format and number of players on heart rate responses and physical demands in small-sided soccer games. Journal of Strength and Conditioning Research. 2012;**27**(5):1295-1303

[30] Clemente F, Martins F, Mendes R. Acute effects of the number of players and scoring method on physiological, physical, and technical performance in small-sided soccer games. Research in Sports Medicine. 2012;**22**(4):380-397

[31] Gaudino P, Alberti G, Iaia F. Estimated metabolic and mechanical demands during different small-sided games in elite soccer players. Human Movement Science. 2014;**36**:123-133

[32] Owen A, Wong D, Paul D, Dellal A. Physical and technical comparisons between various sided games within professional soccer. International Journal of Sports Medicine. 2014;**35**:286-292

[33] Lacome M, Simpson B, Cholley Y, Buchheit M. Locomotor and heart rate responses of floaters during small-sided games in elite soccer players: Effect of pitch size and inclusion of goalkeepers. International Journal of Sports Physiology and Performance. 2108;**13**: 668-671

[34] Sanchez-Sanchez J, Hernández D, Casamichana D, Martínez-Salazar C, Ramirez-Campillo R, Sampaio J. Heart rate, technical performance, and session-RPE in elite youth soccer small-sided games played with wildcard players. The Journal of Strength and Conditioning Research. 2017;**31**(10): 2678-2685

[35] Lozano D, Lampre M, Díez A, Gonzalo-Skok O, Jaén-Carrillo D, Castillo D, et al. Global positioning system analysis of physical demands in small and large-sided games with floaters and official matches in the process of return to play in high level soccer players. Sensors. 2020;**20**:6605

[36] Rábano-Muñoz A, Asian-Clemente J, Sáez de Villarreal E, Nayler J, Requena B. Age-related di erences in the physical and physiological demands during small-sided games with floaters. Sports. 2019;**7**:79

[37] Fradua L, Zubillaga A, Caro O, Fernández-García A, Ruiz-Ruiz C, Tenga A. Designing small-sided games for training tactical aspects in soccer: Extrapolating pitch sizes from full-size professional matches. Journal of Sports Sciences. 2013;**31**(6):573-581

[38] Silva P, Vilar L, Davids K, Araújo D, Garganta J. Sports teams as complex adaptive systems: Manipulating player numbers shapes behaviors during football small-sided games. Springerplus. 2016;**5**:191

[39] Vilar L, Duarte R, Silva P, Chow J, Davids K. The influence of pitch dimensions on performance during small-sided and conditioned soccer games. Journal of Sports Science. 2014; **32**(19):1751-1759

[40] Gonçalves B, Marcelino R, Torres-Ronda L, Torrents C, Sampaio J. Effects of emphasizing opposition and cooperation on collective movement behavior during football small-sided games. Journal of Sports Sciences. 2016; **34**(14):1346-1354

[41] Cruyff J. Me gusta el Fútbol. 1st ed. Barcelona: RBA Libros; 2002

[42] Vilamitjana J, Heinze G, Verde P, Calleja-González J. Comparison of high-intensity patterns between possession games and small-sided games in professional soccer. Acta Gymnica

[43] Asian-Clemente J, Rábano-Muñoz A, Muñoz B, Franco J, Suarez-Arrones L. Can small-side games provide adequate high-speed training in professional soccer? International Journal of Sports Medicine. 2021;**42**(6):523-528

[44] Asian-Clemente A, Rábano-Muñoz A, Requena A, Santalla A, Suarez-Arrones L. The influence of the floater position on the load of soccer players during a 4 vs 4 + 2 game. Kinesiology. 2022;**1**:82-91

[45] Mallo J, Navarro E. Physical load imposed on soccer players during small-sided training games. Journal of Sports Medicine and Physical Fitness. 2008; **48**(2):166

[46] Ade J, Fitzpatrick J, Bradley P. High-intensity efforts in elite soccer matches and associated movement patterns, technical skills and tactical actions. Information for position-specific training drills. Journal of Sports Sciences. 2016;**34**(24):2205-2214

[47] Seirul lo F. El entrenamiento en los deportes de equipo. España: Esteban Sanz; 2017

[48] Hill-Haas S, Coutts A, Dawson B, Rowsell G. Time-motion characteristics and physiological responses of small-sided games in elite youth players: The influence of player number and rule changes. The Journal of Strength and Conditioning Research. 2010;**24**(8): 2149-2156